Process Control Design for Industrial Applications

Series Editor
Bernard Dubuisson

Process Control Design for Industrial Applications

Dumitru Popescu
Amira Gharbi
Dan Stefanoiu
Pierre Borne

iSTE

WILEY

First published 2017 in Great Britain and the United States by ISTE Ltd and John Wiley & Sons, Inc.

ISTE Ltd
27-37 St George's Road
London SW19 4EU
UK

www.iste.co.uk

John Wiley & Sons, Inc.
111 River Street
Hoboken, NJ 07030
USA

www.wiley.com

Library of Congress Control Number: 2017930552

British Library Cataloguing-in-Publication Data
A CIP record for this book is available from the British Library
ISBN 978-1-78630-014-0

Contents

Preface

The purpose of this book is to present the various aspects and the different approaches most commonly employed in the control of industrial processes.

Considering that process control design is carried out using a model based approach, the modeling and identification of the systems are presented with the main objective of producing dynamic control models.

Using the chosen model, the control system is determined so as to ensure that the process satisfies the required level of performance. In the case of linear models, the main methods used in control design are based on the notion of pole placement.

In order to account for the fact that the chosen model is only a simplified and often imperfect description of the process' behavior, more elaborate controls can be suggested: adaptive control, predictive control, internal model control, etc.

When the behavior of the process is strongly nonlinear, the use of a multimodel control can become necessary. The determination, choice and consideration of the various models that can describe the evolution of the process at various operating points depend on the validity of each of these models at the chosen operating points.

We propose a method for estimating the error induced by the models' own estimation difficulties, and by the presence of uncertainties, noise and bounded perturbations.

After presenting the physical laws that govern the evolution of continuous variation processes, we go on to to explore in detail several real optimized control solutions, carried out in an industrial setting, providing the reader with a better understanding of the approaches developed.

Dumitru POPESCU, Amira GHARBI,
Dan STEFANOIU and Pierre BORNE
February 2017

List of Notations and Acronyms

$(\hat{a}(z^{-1}), \hat{b}(z^{-1}))$	Dynamic control model
$(\hat{a}_m(z^{-1}), \hat{b}_m(z^{-1}))$	Dynamic tracking model
AF-CLOE	Adaptively Filtered Closed Loop Output Error (identification method)
A,B,C,D	State-space representation of the continuous MIMO system
A_d,B_d,C_d,D_d	State-space representation of the discrete MIMO system
A,b,c,d	State-space representation of the continuous SISO system
A_d,b_d,c_d,d_d	State-space representation of the discrete SISO system
ARMAX	Model or class of identification models expressed by 3 terms: autoregressive (AR), moving average (MA) and exogenous control (X)
ARX	Identification model of autoregressive type (AR), with exogenous control (X)
(C,M)	Closed loop nominal system
(C,P)	Closed loop real system
DPRC	Differential Pressure Control System
FRC	Flow Control System
LRC	Level Control System
LS	Least Squares identification technique
RLS	Recursive Least Squares identification technique

PID	Proportional-integral-derivative algorithm
PRC	Pressure Control System
SM	State Model
TRC	Temperature Control System
BJ	Identification model of Box-Jenkins type
CL	Closed Loop (system, identification method etc.)
CLOE	Closed Loop Output Error (idenfication methods)
CLSI	Closed Loop System Identification
dB	decibel(s) – measuring unit for the signals/systems spectra
E-LSM	Extended Least Squares Method
F-CLOE	Filtered Closed Loop Output Error (identification method)
FIR	Finite Impulse Response (filter, system)
FT	Fourier Transform
G-CLOE	Generalized Closed Loop Output Error (identification method that replaces ARX model by BJ model)
G-LS	Generalized Least Squares (PEMM for the BJ model)
$G(s)$	Continuous system transfer function
$G(z)$	Discrete system transfer function
$G_R(z^{-1})$, $G_S(z^{-1})$	Pre-specified polynomials for robust control
I-CLOE	Integral Closed Loop Output Error (identification method)I/O Input-Output (type of identification model, transformation, operator, etc.)
IIR	Infinite Impulse Response (filter, system)
I=f(V)	Photovoltaic Current-Voltage characteristic
L	Estimator matrix
LSM	Least Squares Method
M	Sylvester matrix
MIMO	Multi-Input Multi-Output (type of fully multi-variable model or system or process)

MISO	Multi-Input Single-Output (type of multi-variable model or system or process with several inputs and on single output)
MV-LSM	Multi-Variable Least Squares Method
OL	Open Loop (system, identification etc.)
OLOE	Open Loop Output Error (identification method)
OLSI	Open Loop System Identification
PEMM	Prediction Error Minimization Method (identification method)
$P(z^{-1})$	Characteristic polynomial of the system
P=f(I,V)	Photovoltaic Power-Current, Voltage characterstic
PRS	Pseudo-Random signal
PV	Photovoltaic pannel
Q	Observability matrix
R	Controlability matrix
R-ELS	Recursive Extended Least Squares (identification method)
RST	Automatic regulator with 3 polynomials: R (regulation), S (sensitivity) and T (tracking)
RST-YK	RST regulator expressed in Youla-Kucera parametric form
SI	System identification
SISO	Single-Input Single-Output (type of model or system or process with one input and one output)
SNR	Signal-to-Noise Ratio
$S_{vy}(j\omega)$	Disturbance-output sensitivity function
W-CLOE	Weighted Closed Loop Output Error (identification method)
X-CLOE	Extended Closed Loop Output Error (identification method that replaces ARX model with ARMAX model)
X-OLOE	Extended Open Loop Output Error (identification method employed in case of ARMAX model instead of ARX model)
YK	Youla-Kucera (parametric expressions of a regulator)
$\|\Delta M(j\omega)\|$	Modulus margin of the system robustness

1

Introduction – Models and Dynamic Systems

This chapter is a presentation of the issues surrounding the design and implementation of automatic control solutions for industrial applications. Currently, there is an increasing demand for setups that are productive and offer a good yield for the provider by means of modern concepts such as "embedded optimization" and "low-cost automation". In this context, Automation and Applied Informatics, through their theoretical tools, as well as hardware and software resources, contribute toward the development of applications using the existing technologies as well as the emergent ones. Automation makes it possible to describe the evolution of a real process using an abstract mathematical model, and to introduce the necessary methods and mechanisms to design the advanced control systems that ensure an optimal operation of the process. On the other hand, Informatics, due to the hardware material and software tools available and the efficiency of its powerful computing and communication which contribute toward operations involving data processing, model development, design and identification of control methods, is vital for the optimization, surveillance and protection of industrial processes. The different classes of dynamic models (systems) associated with the processes, as well as the most important algorithms, methods and techniques used in the design of the control systems for industrial facilities, are briefly presented here.

1.1. Overview

The design and implementation of digital control systems for industrial applications has, as a primary objective, the valorization of Automatic Control and Informatics tools in order to improve the quality of products, to ensure the safety of personnel, protect industrial facilities and the environment, and decrease as much as possible the production costs.

This objective is a response to the increasing demands in efficiency and autonomous operating requirements of companies, the need to reduce the role of the operator during process use (facilities, machines and equipment start to communicate with each other using the mechanisms involved in the concept of the *internet of things*) and the augmenting integration of automation into high-performance production technologies.

Two main categories of Industrial Automatic Control that are required are as follows: first, the mathematical formulism that allows the process to be represented by a mathematical model that expresses its dynamics, and second, the concepts and methods used in the control and optimization of the process [OGA 90, KUO 91, AST 97, GEN 97, POP 06].

A mathematical model is set using the laws that govern the functioning of the process or by using a collection of data acquired from the process.

The control algorithms are calculated based on the model, using numerical methods so as to ensure the level of performance desired in terms of the real processes. In the following step, which has become mandatory, an optimal operating point is found in the real operating conditions using an adequate decision criterion for the process.

The digital methods for computing and programming also ensure acquisition and processing of the data in real time and assist the design of control systems, as well as aiding the control and surveillance of industrial applications.

The classic control solution ensures a good response around a nominal operating point. To achieve a more widespread performance around the nominal point, this type of control must be improved using adaptive, robust, predictive, multimodel or intelligent techniques.

Normally, the exploitation of the process takes place within an "admissibility domain"; this operating domain of the process is imposed by technological limitations.

A standard control process configuration is organized in several steps over two important automation levels: the control (execution) level and the supervisory (decision) level. The supervisory level calculates, based on logical user demands, the optimal decision that can be implemented effectively by the action of the control systems at the execution level. In terms of regulation, the process evolves around the nominal operating point (NP) under the effect of the advanced control strategy, while conserving performance, in a prespecified vicinity of the NP.

The supervisory system finds the optimal operating point (OP) in the admissibility domain for the suitable optimization techniques, and the decision of the supervisor is transferred automatically from NP to OP. Sometimes, the case arises in which the process leaves the admissible operating domain, at the fault point, in which case the abnormal situation is detected and diagnosed so as to eliminate the cause of the fault as soon as possible. After the reconfiguration of the control systems, so that they become more tolerant to faults, the process operates in a normal state [POP 06, GEV 95, BOR 93].

In this context, our work, located in the control level area, proposes the necessary ingredients for the design and implementation of modern control systems as part of industrial applications and is notably directed toward systems engineering specialists.

We must note the complementarity that exists between this work and two books on optimization in engineering sciences [BOR 13, STE 14], which explain the methodology and techniques required for the implementation of the supervisory level in high-performing solutions of control in industrial processes.

1.2. Industrial process modeling

A mathematical model is the abstract representation of the behavior of a process (in this case, industrial). In other words, it is a dynamic system associated with the process that describes its behavior in a simple manner and allows a better understanding of the examined phenomena in order to carry out simulations, thus avoiding the need to conduct experiments that could be costly or get in the way of the process.

The model obtained will always be at best an idealization of the physical reality, considering that a hypothesis had to be made to obtain it. Furthermore, the search for a model can take on various forms, depending on the objective in question, the tools available and the nature and complexity of the phenomenon or process involved. In the engineering sciences domain, at the origin of the estimation of a dynamic model there is a set of equations that uses the conservation principle relative to certain physical generalized quantities: matter, mass, energy, amount of movement and electric charge. This mechanism of model estimation is reflected by the creation of balance equations for each of these quantities when they are implicated [DAU 04, POP 06, POP 11].

Most often, we start by writing the balance equation that expresses the preservation of a quantity W (mass or energy) over a time dt. The general form of this type of equation is the following:

(Variation dW of W in the system) = (entering quantity of W in the system) – (exiting quantity of W in the system) + (accumulated quantity of W) – (consumed quantity of W). For example, in the case of a representative chemical process, this can be written as

$$dW = Q_e dt - Q_s dt + r_f dt - r_c dt \qquad [1.1]$$

or again,

$$\frac{dW}{dt} = Q_e - Q_s + r_f - r_c, \qquad [1.2]$$

where r_f and r_c represent, respectively, the direct and reverse reaction rates for the transformation of the value W in the system.

The conservation principle of the global mass of the system can be written as

$$\frac{dM}{dt} = \frac{d(\rho V)}{dt} = \rho_e Q_e - \rho_s Q_s, \qquad [1.3]$$

where ρ is the density of the matter in the system, ρ_e is the density of the matter in the entering flow, ρ_s is the density of the matter in the exiting flow, Q_e is the entering flow rate, Q_s is the exiting flow rate, V is the volume of the system and M is the total mass of the system.

For the conservation of each component, it is known that

$$\frac{dn_a}{dt} = \frac{d(c_a V)}{dt} = c_{a,in} Q_e - c_{a,out} Q_s + (r_{af} - r_{ac})V, \qquad [1.4]$$

where c_a is the molar concentration of component A in the system, $c_{a,in}$ is the molar concentration of component A in the entering flow, $c_{a,out}$ is the molar concentration of component A in the exiting flow, Q_e is the entering flow rate, Q_s is the exiting flow rate, V is the volume of the system, n_a is the number of moles in the system and r_{af}, r_{ac} are the reaction rates of component A in the direct and reverse direction of the reaction.

For the conservation of energy:

$$\frac{dE}{dt} = \frac{d(U+K+P)}{dt} = h_e \rho_e Q_e - h_s \rho_s Q_s \pm Q_c \pm F_s, \qquad [1.5]$$

where E is the total energy of the system, U is the internal energy, K is the kinetic energy, P is the potential energy, h_e is the specific enthalpy of the matter in the entering flow, h_s is the specific enthalpy of the matter in the exiting flow, Q_c is the quantity of heat exchanged between the system and the environment and F_s is the action of the external forces acting on the system.

For the more complex processes, there is also the possibility of a synthetic representation of the main characteristics of the process from equations constructed in a systemic way to a certain degree of precision.

There are two important classes of models used in Automatic Control: state space models (SSMs) based on the structural state concept, and the input–output models (IOMs) based on the concept of causality.

Depending on the nature of the process and on the goal of the user, the balance equations can be easily transformed into a SSM or into an IOM, which express the dynamics of the process [AST 93, CAL 79, KAI 91, KAP 91, COR 13, DAU 04, POP 06, POP 11].

1.3. Model classes

Here, we present the most common models used in the design of control systems: SSMs and IOMs.

1.3.1. *State space models*

SSMs are connected to the structural transformations of a process that manipulates internal variables used in the study of dynamic systems. Let us consider a physical process with an entering quantity of W_i (primary matter) and an exiting quantity W_e (product), which is the result of a set of transformations of mass and/or energy.

In the stationary operating regime of the process, the two quantities are equal to each other:

$$W_{e0} - W_{i0} = 0 \qquad [1.6]$$

a fact that, through the mass or energy balance, explains the state of equilibrium of the process.

Seeing as the consumer's demands can vary, the exiting quantity W_e is variable in time and the state of the process is therefore unbalanced; this results in a dynamic state. Equilibrium returns through the action of the variables that produce accumulation phenomena in the process in question. Let us consider the internal variables of the process gathered in the vector of the state variables *x(t)*, so relation [1.6] becomes

$$\dot{x}(t) = W_i(t) - W_e(t)\,. \tag{1.7}$$

The near majority of industrial processes are self-stabilizing and therefore an extra negative term a is introduced into relation [1.7], and thus it becomes

$$\dot{x}(t) = ax(t) + W_i(t) - W_e(t)\,. \tag{1.8}$$

By a simple generalization, considering *x(t)* to be a state vector, relation [1.8] becomes

$$\dot{x}(t) = Ax(t) + W_i(t) - W_e(t)\,, \tag{1.9}$$

where A is the state matrix that normally ensures the stabilizing character of the system through its eigen values with real negative parts.
The change in the entering quantity W_i is considered by the control vector u that influences the state variables using weighting elements included in a matrix B, and relation [1.9] is written as

$$\dot{x} = Ax + Bu - W_e\,. \tag{1.10}$$

It is reasonable to consider that the output values gathered into the vector y are determined from state variables or are sometimes directly influenced by the control vector u *via* matrix D [CAL 79, FIL 02].

In this case, the process considered is expressed in general through the following state dynamics model:

$$\begin{aligned} \dot{x}(t) &= Ax(t) + Bu(t) \\ y(t) &= Cx(t) + Du(t) \end{aligned}\,. \tag{1.11}$$

The set (A, B, C, D) characterizes the state space representation of the process, where $u(t): \mathbb{R} \to \mathbb{R}^l$, $x(t): \mathbb{R} \to \mathbb{R}^n$ and $y(t): \mathbb{R} \to \mathbb{R}^p$.

The representation in [1.10] is a multivariable Multi-Input Multi-Output (MIMO) continuous, linear and non-varying system.

For a monovariable SISO system, relation [1.10] is written as

$$\begin{aligned} \dot{x}(t) &= Ax(t) + bu(t) \\ y(t) &= c^T x(t) + du(t) \end{aligned}, \qquad [1.12]$$

where b and c are vectors in $\mathbb{R}^n$ and d is a scalar.

The system:

$$\begin{aligned} \dot{x}(t) &= A(t)x + B(t)u \\ y(t) &= C(t)x + D(t)u \end{aligned}, \qquad [1.13]$$

where the matrices A, B, C, D are dependent on time is a continuous, linear and varying system.

The system that is represented by the relations in [1.14] is a nonlinear system:

$$\begin{aligned} \dot{x} &= f(t, x, u) \\ y &= g(t, x) \end{aligned}, \qquad [1.14]$$

where functions f and g are, respectively, made up of n and p components,

$$f : \mathbb{R}x\mathbb{R}^n \rightarrow \mathbb{R}^n, \; g : \mathbb{R}x\mathbb{R}^n \rightarrow \mathbb{R}^p. \qquad [1.15]$$

The equivalent discrete models are of the following form:

$$\begin{aligned} x(k+1) &= A_d x(k) + B_d u(k) \\ y(k) &= C_d x(k) + D_d u(k) \end{aligned} \qquad [1.16]$$

for the linear, non-varying MIMO system and

$$\begin{aligned} x(k+1) &= A_d x(k) + b_d u(k) \\ y(k) &= c_d x(k) + d_d u(k) \end{aligned} \qquad [1.17]$$

for the SISO representation.

The nonlinear discrete model becomes

$$\begin{aligned} x(k+1) &= f(k, x, u) \\ y(k) &= g(k, x) \end{aligned}. \qquad [1.18]$$

In the usual case of linear, non-varying, continuous systems [1.13], the state trajectory is obtained as the unique solution

$$x(t) = e^{A(t)}x(0) + \int e^{A(t-\tau)}Bu(\tau)d\tau \qquad [1.19]$$

with $F(t)$ the continuous function and $F(k)$ the discrete function, respectively.

$$F(t) = e^{At}, t \in \mathbb{R} \quad F(k) = e^{k}, k \in \mathbb{Z} \qquad [1.20]$$

We denote h as the sampling period.

There are known relations between the sets of (A, B, C, D) and (A_d, B_d, C_d, D_d), and for the monovariable system, for example, these correspond to the following expressions:

$$A_d = e^{Ah}; B_d = (\int_0^h e^{A\tau}d\tau)b; c_d = c^T; d_d = d. \qquad [1.21]$$

In this case, it is easy to evaluate the continuous system response (weighting function) of the Dirac impulse (A, b, c, d)

$$g(t) = c^T F(t)b, t \in \mathbb{R} \qquad [1.22]$$

and for the discrete system,

$$g(k) = c_d^T F(k)b_d, k \in \mathbb{Z}. \qquad [1.23]$$

Next, we have looked at the tangent models defined around an operating point, and therefore at non-varying linear systems [CAL 79, KUO 91, WIR 91, DOI 97, DAU 04].

In this case, we can study the relationship between the two system representations of the process models. (A, B, C, D) is the non-varying linear system and $\nu(A)$ is the spectrum of matrix A (the eigen values of matrix A):

$$\nu(A) = \{\nu_1, \nu_2, \ldots, \nu_n\}. \qquad [1.24]$$

The norm of the fundamental matrix (state transition matrix) verifies the following relation:

$$\|F(t)\| = \|e^{At}\| < Me^{at}, a = \max\{\mathrm{Re}(v_i), i = \overline{1,n}\}. \qquad [1.25]$$

By applying the Laplace operator, we obtain the transfer matrix of the continuous system:

$$G(s) = C(sI - A)^{-1}B \qquad [1.26]$$

which is a rational fractions matrix with real coefficients:

$$G(s) = [g_{ij}(s)] = \left[\frac{D_{ij}(s)}{E_{ij}(s)}\right]. \qquad [1.27]$$

We must note that all rational fractions $g_{ij}(s)$ have the same denominator and are strictly proper (the order of $D(s)$ is lower than the order of $E(s)$).

For the discrete system, the transfer matrix is calculated in the same way:

$$G_d(z) = C_d(zI - A)^{-1}B_d \qquad [1.28]$$

with irreducible rational fraction elements that are strictly proper:

$$G_d(z) = [g_{ij}(z)] = \left[\frac{D_{ij}(z)}{E_{ij}(z)}\right]. \qquad [1.29]$$

The result in the monovariable case is obvious for a continuous system

$$G(s) = c^T(sI - A)^{-1}b \qquad [1.30]$$

and for the equivalent discrete system

$$G(z) = c_d^T(zI - A_d)^{-1}b_d. \qquad [1.31]$$

Some structural properties of non-varying linear systems are presented below.

A continuous system is asymptotically stable (internal stability) if the spectrum of the matrix $V(A)$ is located in $\mathbb{C}^-$.

A discrete system $\dot{x} = A_d x$ is asymptotically stable if all of the eigen values have a modulus of less than 1, meaning that the spectrum of the matrix $V(A_d)$ is located in the unit circle $C(0,1) = (z \in \mathbb{C}, |z| \leq 1)$.

Consider, for example, a monovariable continuous system,

$$\begin{aligned} \dot{x}(t) &= Ax(t) + bu(t) \\ y(t) &= c(t)x(t) + du(t) \end{aligned}, \quad [1.32]$$

where the matrix A is of rank n.

The controllability of the system targets the couple (A, b), and therefore the SISO system [1.32] is controllable if the controllability matrix $R = (b, Ab, A^2b, A^3b, ..., A^{n-1}b)$ has a rank of n.

In an equivalent way, a MIMO system is always controllable if matrix $R = (B, AB, A^2B, A^3B, ..., A^{n-1}B)$ has the rank n.

For a controllable system, there is a control algorithm that can change the dynamics of the systems through state transitions. Controllability is linked to the property of the stabilizability of systems, providing strategies for stabilizing an unstable system, and ensuring desired levels of performance, using a state feedback control.

For a continuous system with A of rank n, the realization (A, B, C) must therefore be minimal (the dimension of the state space is n). This condition is reflected in the case of MIMO systems through the transfer matrix:

$$G(s) = [g_{ij}(s)] = \left[\frac{N_{ij}(s)}{D_{ij}(s)} \right], \deg(N_{ij}(s)) < \deg(D_{ij}(s)) = n \quad [1.33]$$

which must contain real elements and irreducible rational fractions. In the case of SISO systems, this condition is reflected in the transfer function,

$$G(s) = \left[\frac{N(s)}{D(s)} \right], \deg(N(s)) \leq \deg(D(s)) \quad [1.34]$$

which is an irreducible rational fraction.

The most important consequence that exploits the quality of a minimal realization of the system (A,b,c,d) is to calculate the SSM, which corresponds to the I/O model.

Consider, for example, the transfer function of a SISO system, expressed by the irreducible fraction,

$$G(s) = \frac{b_n s^n + b_{n-1} s^{n-1} + \ldots + b_0}{a_n s^n + a_{n-1} s^{n-1} + \ldots + a_0} \qquad [1.35]$$

and so the corresponding minimal realization is the following:

$$A = \begin{bmatrix} 0 & 1 & 0 \ldots 0 \\ 0 & 0 & 1 \ldots 0 \\ -a_0 & -a_1 & \ldots\ldots -a_n \end{bmatrix},$$

$$b^T = (0 \quad 1 \quad 0 \quad 00 \quad 0 \quad 0 \quad 0),$$

$$c^T = (b_0 - b_n a_0, b_1 - b_n a_1, \ldots, b_{n-1} - b_n a_{n-1}), \qquad [1.36]$$

$$d = b_n$$

The observability property targets the couple (A,C) and the system [1.12] is observable if the matrix $Q^T = (c, Ac, A^2c, A^3c, \ldots, A^{n-1}c)$ has a rank of n for SISO systems, and equivalently, if the observability matrix $Q^T = (C, AC, A^2C, A^3C, \ldots, A^{n-1}C)$ has a rank of n, for MIMO systems.

Observability is linked to the property of detectability; the detectability of the set (A,C) provides the possibility to construct a stable state estimator associated with the original system and notably to find a matrix L for which the spectrum $V(A+LC)$ is located in the domain $\mathbb{C}^-$ for continuous systems, and $V(A_d + LC_d)$ is located in the circle $C(0,1)$ for discrete systems.

The dynamics of the estimator is described in the following equations:

$$\dot{\hat{x}} = (A + LC)\hat{x} + Bu \,, \qquad [1.37]$$

where $\hat{x}(t)$ is the state vector estimated using the observations made over $u(t)$ and $y(t)$.

The SSMs are used in designing the state feedback control. The control *u(t)* is computed through the relation,

$$u(t) = -Kx(t)\,, \qquad [1.38]$$

where matrix K is determined by a pole placement method.

In the case of observer-based control, it is computed in a similar manner:

$$u(t) = -K_e\hat{x}(t) \qquad [1.39]$$

The optimal control uses the integral criterion:

$$I = \int_0^{\infty}\left[x^T(t)Qx(t) + u^T(t)Ru(t)\right]dt\,, \qquad [1.40]$$

$$Q \geq 0, R > 0$$

where matrices Q and R (symmetric positive-definite) are chosen to ensure the supplementary qualities of the control *u(t)*, which minimizes criterion [1.40] and is expressed as follows:

$$u(t) = -R^{-1}B^T Px(t)\,. \qquad [1.41]$$

In expression [1.41], matrix P, which is symmetric positive-definite, is the solution of the matricial algebraic Riccati's equation:

$$Q + A^T P + PA - PBR^{-1}B^T P = 0\,. \qquad [1.42]$$

For the integrating term from [1.40], we get

$$x^T(t)Qx(t) + u^T(t)Ru(t) \qquad [1.43]$$

which is positive-definite, the expression of matrix K can be calculated employing Lyapunov functions based techniques [CAL 79, KUO 91, WIR 91, DIO 97, DOI 97, DAU 04].

1.3.2. *Input–output models*

The design of IOM is based on mass or energy balance equations, respecting the principle of causality. In order to reach the form needed in the control design, we start with the nonlinear knowledge-based model, which expresses the operating laws of the process. We continue with the tangential and normalized behavioral model and finally we introduce the control models in the design of the control systems. All these different classes of models are presented later [DAU 04, POP 06, BOR 10].

1.3.2.1. *Knowledge-based model*

This type of model is usually established from the knowledge of the internal laws, which govern the evolution of the process. Their transcription leads to one (or several) analytical expression(s) of a model whose parameters have physical significance.

In process engineering studies, there are some fundamental principles of classical physics: the principle of the conservation of mass, the fundamental principle of mechanics (conservation of the quantity of movement), the first principle of thermodynamics (conservation of energy) and the second principle of thermodynamics (conservation of entropy).

When chemical, biochemical or microbiological transformations are at play, the kinetics of the reactions must be looked at and knowledge of these is vital in establishing a reactor model. Therefore, knowledge-based models mathematically express the phenomena responsible for the process dynamics.

The first phase in obtaining a knowledge-based model involves defining the system and the phenomena that are to be represented within it. The limits of the physical system must be determined, as must the variables that permit the representation and identification of important physical, chemical or biological phenomena (see section 1.2).

The main objective of "knowledge-based" models is, therefore, to describe a phenomenon using a mathematical relation. They remain very close to the real processes enabling a reproduction of the behavior of the process over a large operating domain. In general they are quite complex, with a large number of parameters which characterize the nonlinearity of the process expressed as technological variables [TER 90, FOU 04].

1.3.2.2. *Behavioral model*

A behavioral model is a systemic representation that uses differential equations or transfer functions to help simulate the dynamics of the process. Generally speaking, a behavioral model is expressed in a standard format and its coefficients have physical significance.

The behavioral model can be obtained following a few mathematical transformations carried out on the raw version of the knowledge-based model [DAU 04].

Using inductive reasoning, it can be stated that, around an operating point, small variations in the input of a system can be linked to small variations in the output for a linear model, which is expressed as:

$$\begin{aligned} &a_n \frac{d^n y(t)}{dt^n} + a_{n-1} \frac{d^{n-1} y(t)}{dt^{n-1}} + \ldots + a_1 \frac{dy(t)}{dt} + a_0 y(t) \\ &= b_m \frac{d^m u(t)}{dt^m} + b_{m-1} \frac{d^{m-1} u(t)}{dt^{m-1}} + \ldots + b_1 \frac{du(t)}{dt} + b_0 u(t) \end{aligned} \qquad [1.44]$$

where coefficient n specifies the order of the system.

This equation is linear with constant coefficients, and the input–output dynamic behavior of the system can be represented simply with a mathematical tool. In order to easily handle relation [1.44], the Laplace transform is used, which enables the passage from a differential equation to an algebraic one.

Here, we shall evaluate the Laplace transform of equation [1.44]. First, we detail the initial conditions: the value of $y(0)$ and of all its derivatives are given, and the value of $u(0)$ is zero (zero action before $t = 0$).

The Laplace transform and its properties can then be used in order to obtain the relation:

$$\begin{aligned} &(a_n s^n + a_{n-1} s^{n-1} + \ldots + a_0) Y(s) - \Big[s^{n-1} (a_n y(0)) \\ &+ s^{n-2} \left(a_n \frac{dy}{dt}(0) + a_{n-1} y(0) \right) + \ldots \\ &+ s^0 \left(a_n \frac{d^{n-1} y}{dt^{n-1}}(0) + \ldots + a_1 y(0) \right) \Big] = (b_m s^m + b_{m-1} s^{m-1} + \ldots + b_0) U(s) \end{aligned} \qquad [1.45]$$

The expression in square brackets corresponds to the free response of the system, meaning the response that represents the effect of the initial conditions. The other term, as a factor of $U(s)$, corresponds to the forced response, meaning the response to the input, where the transfer function is defined by considering the case where all of the initial conditions are equal to zero:

$$G(s) = \frac{b_m s^m + b_{m-1} s^{m-1} + \ldots + b_0}{a_n s^n + a_{n-1} s^{n-1} + \ldots + a_0}. \qquad [1.46]$$

Equivalent models exist for discrete systems expressed by the difference equation:

$$\begin{aligned}&a_0y(k)+a_1y(k-1)+...+a_ny(k-n)=\\&b_0u(k)+b_1u(k-1)+...+b_mu(k-m)\end{aligned}\qquad [1.47]$$

and, respectively, by a transfer function in z^{-1}:

$$G(z^{-1})=\frac{Y(z^{-1})}{U(z^{-1})}=\frac{b_0+b_1z^{-1}+...+b_mz^{-m}}{a_0+a_1z^{-1}+...+a_nz^{-n}}. \qquad [1.48]$$

It must be noted that generally the term b_0 is equal to zero (no direct transfer of information).

It is clear that a behavioral model does a good job of expressing the structural component of the process, as the differential equation or the transfer function expresses the order of the mathematical model. This type of model offers the possibility to propose an adequate structure for a control model, which is evaluated through process identification.

The behavioral model is then based on the knowledge-based model and used in the evaluation of the dynamics of a real process through simulation. This type of model provides information regarding the structure of the model through a standard differential equation in which the coefficients are normalized.

The behavioral model is especially interesting, on the one hand, through its relation with knowledge-based models and, on the other hand, for its link with control models [DAU 04, POP 11].

The control model is used in the designing of closed-loop systems.

1.3.2.3. *Control model*

The control model is one of the dynamic models, which are obtained through a particular representation of a behavioral model, expressed using parameters usually handled in Automatic Control (gain, time constant, pure delay, etc.) and which must be identified by the process specialist, for the control design of digital systems. The availability of digital resources enables the establishment of automatic estimation algorithms for the parameters of a model through an identification operation. It is important to emphasize that the identification operation can provide more general, high-performing discrete models and holds several advantages over other approaches.

High-performance identification algorithms have been extensively developed as a result of their recursive implementation adapted to real-time identification problems and their application to numerical systems.

The estimation principle for the parameters of the sampled models is based on the prediction error of the model.

An adjustable parameter sampled model is easily implanted into numerical systems. The error between the output of the process $y(t)$ and the output $\hat{y}(t)$ predicted by the model, called the prediction error, is used by a parametric adaptation algorithm, which at each sampling moment modifies the parameters of the model so as to minimize this error. Generally, the output is a low-level pseudo-random binary sequence produced by the computer (a succession of rectangular pulses of randomly variable duration). Once the model is obtained, an objective validation can be carried out using statistical tests regarding the prediction error $\varepsilon(t)$. The validation test makes it possible, for a given process, to choose the best algorithm for estimating the parameters and the best model. This direct approach for the identification of dynamic process models has a range of other advantages, such as:

– significant data compression, as at any given moment the recursive algorithms only deal with an input/output pair rather than with all of the input/output data;

– relatively low memory and computational power requirements;

– easy installation onto a microcomputer;

– possible realization of identification systems in real time;

– possibility of tracking slowly varying parameters over time;

– identification of measurable perturbation models.

One of the key elements in the establishment of this approach for the identification of process models is the *parameter adaptation algorithm*, which drives the parameters of the adjustable model at each sampling step. This algorithm has a "recursive" structure, meaning that the new value of the parameters is equal to the most recent value plus a correction term, which depends on the most recent measurements. Usually, a "vector" of the parameters is defined and its components are the different parameters to be identified. Parameter adaptation algorithms have the following structure:

[New estimation of the parameters (vector)] = [Previous estimation of the parameters (vector)] + [Adaptation gain (matrix)] × [Function of the measurement (vector)] × [Function of the prediction error (scalar)]. The vector of the functions of the measurements is called the "*vector of the observations*".

We consider the model in discrete time:

$$\hat{y}(k+1)=\sum_{i=1}^{nA}\hat{a}_i y(k+1-i)+\sum_{i=1}^{nB}\hat{b}_i u(k+1-i) \qquad [1.49]$$

and using the notations

$$\hat{\theta}^T(k+1)=\left[\hat{a}_1,\hat{a}_2,\ldots,\hat{a}_{nA},\hat{b}_1,\hat{b}_2,\ldots\hat{b}_{nB}\right] \qquad [1.50]$$

$$\Phi^T(k)=\left[-y(k),-y(k-1),\ldots,-y(k+1-nA);u(k),\ldots,u(k+1-nB)\right] \qquad [1.51]$$

we obtain

$$\hat{y}(k+1)=\hat{\theta}^T(k)\Phi(k),\forall k\in\mathbb{Z}\,. \qquad [1.52]$$

We use a suitable identification method, for example Recursive Least Square (RLS), to estimate the parameters of the model as the solution of the optimization problem:

$$\min_{\hat{\theta}(k)}\{J(\hat{\theta}(k))=\sum_{i=1}^{k}\left[y(i)-\hat{\theta}^T(k)\Phi(i-1)\right]\} \qquad [1.53]$$

The solution to problem [1.53] is given in the following recursive relation:

$$\hat{\theta}(k+1)=\hat{\theta}(k)+F(k+1)\Phi(k)\varepsilon(k+1) \qquad [1.54]$$

with

$$F(0)=\alpha,I(\alpha>0);\hat{\theta}(0)=\theta_0 \qquad [1.55]$$

and

$$F(k+1)<F(k)\,. \qquad [1.56]$$

The recursive algorithm [1.54] ensures the convergence of the RLS method, and the estimation error is reduced to zero, asymptotically,

$$\lim_{k\to\infty}\varepsilon(k+1)=0\,. \qquad [1.57]$$

From equation [1.49], we easily obtain the polynomial form of the model:

$$\hat{y}(z^{-1}) = \frac{\hat{b}(z^{-1})}{\hat{a}(z^{-1})} u(z^{-1}) \qquad [1.58]$$

where the set

$$\begin{aligned} \hat{a} &= 1 + a_1 z^{-1} + \ldots + a_{nA} z^{-nA} \\ \hat{b} &= b_1 z^{-1} + b_2 z^{-2} + \ldots + b_{nB} z^{-nB} \end{aligned} \qquad [1.59]$$

is considered to be the model used for the control design of digital systems [LAN 93, LAN 97, POP 00, POP 01, DAU 04].

For the evaluation of knowledge-based behavioral and control models, linearization around an operating point (tangent systems) for different types of industrial processes is necessary in order to reach the step of designing the system control algorithm in the closed-loop system.

We can cite [LAN 93, GEN 97, POP 06, BOR 11] and simply appeal to the design of model-based systems using the method of pole placement. We consider the structure of the looped system in Figure 1.1, where $(\hat{b}(z^{-1}) / \hat{a}(z^{-1}))$ is the control model, and the set ($R(z^{-1}), S(z^{-1})$) is the control algorithm.

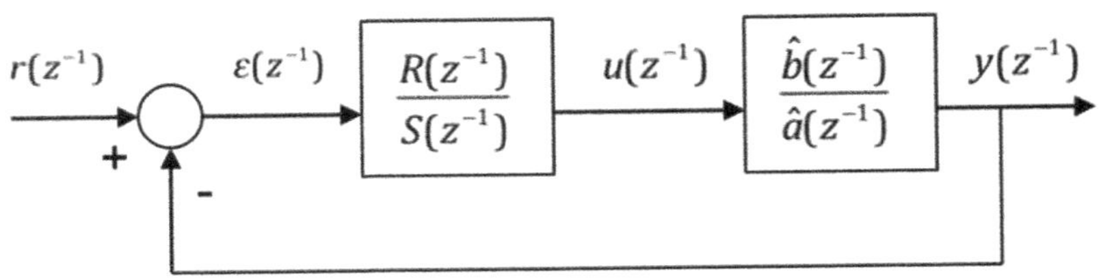

Figure 1.1. *Digital system in a closed-loop system*

We note $P_i(z^{-1})$ the characteristic polynomial that imposes the level of performance. The solution for the Diophantine polynomial equation:

$$P_i\left(z^{-1}\right) \stackrel{\Delta}{=} \hat{a}\left(z^{-1}\right) S\left(z^{-1}\right) + \hat{b}\left(z^{-1}\right) R\left(z^{-1}\right)$$

provides the unknown polynomials of the controller ($R^*(z^{-1}), S^*(z^{-1})$). This digital control can be improved following adaptive, robust, predictive or multimodel strategies [LAN 97, DAU 04, POP 06, BOR 11].

1.3.2.4. *Decision models*

In this case, the process is considered to be a "black box", and the establishment of the model is based on observations of the inputs/outputs of the process. The inputs of the model are the parameters controlled by regulation and the outputs are considered to be quality indicators. The model obtained must appropriately describe the behavior of the process in its near-stationary operating mode. This class of models mainly involves complex, nonlinear and multivariable industrial processes and is used in the resolution of optimization problems. The solution to these problems represents the optimal decision for the process, introduced as set points for the closed-loop systems, ensuring the optimal operating point. The structure of the decision models is inspired from laws that govern the process, or from a sensitivity analysis of the outputs of the process in relation to the variation of the inputs. The parameters of the model are estimated using experimental techniques (MC-off line), following a protocol of data acquisition and processing. The models are checked and updated whenever it is necessary [CAL 79, TER 00, FIL 09, LUP 13].

This procedure starts with the programming of the experimental data acquisition from the process. After a data processing protocol that chooses the inputs and the output(s) of the model, the structure of the multivariable nonlinear model is stated and the data files are prepared for the identification operation itself. The output of the model represents the order of quality $\hat{z}$ of the process in question and the inputs represent the controlled variables. The general representation of the model is therefore a nonlinear function of unknown parameters $\hat{a}$ and the input vector y:

$$\hat{z} = f(\hat{a}, y), \tag{1.60}$$

where the estimator $\hat{a}$ of the parameters of the model is calculated using the least squares method over the whole of the experimental dataset. If the data are organized in a matricial–vectorial manner, the estimator $\hat{a}$ is calculated following the relation:

$$\hat{a} = \left(YY^T\right)^{-1} Y^T z . \tag{1.61}$$

Here, Y is the matrix of the observations acquired at the inputs and z is the vector of the observation acquired at the output.

Using an algebraic operator Q, the decision model and the associated constraints are used for building the optimization problem for the process in question:

$$\min\{I(y) = Q[\hat{z}(y)]\}. \tag{1.62}$$

In its first part, this work presents the concepts, the methods and the highest performing techniques used in the design of digital systems to specialists in Industrial Automatic Control. In the final part, it presents examples for the implementation of practical applications in the use of industrial processes [CAL 79, TER 98, POP 06].

In Chapter 2, we present the methods of identification (in the closed-loop system) that provide the best levels of performance in the estimation of the digital form of dynamic process models, which are necessary for the design of control systems.

2

Linear Identification of Closed-Loop Systems

This chapter introduces some closed-loop (CL) systems linear identification methods, taking the R (regulation), S (sensitivity) and T (tracking) (RST) type regulator as an example of an automatic controller. It concludes with an application of CL identification, in which another regulator is used (of the Youla–Koucera type).

2.1. Overview of system identification

One of the pillars supporting the Automatic Control domain is the field of *System Identification* (SI). This field includes quite a large set of techniques to define and estimate mathematical models associated with real systems/processes (whether they are linear or nonlinear). Usually, such models are digital and can be estimated by using measured data provided by the system/process under discussion. The Automatic Control and SI were actually developed at approximately the same time (at the end of the 1930s). They have been progressively developed alongside each other, as, traditionally, each identification model allowed for the creation of an automatic control law for the corresponding process.

Overall, there are two categories of SI techniques: for processes without automatic regulation (*open loop* [OL]) and for processes containing an integrated automatic controller (*closed loop* [CL]).

The main goal of *open-loop SI* (OLSI) is to provide a sufficiently accurate digital model from which an appropriate regulator can be designed. A range of linear identification methods in an OL is described in [SCS 05, BOR 13a] or [BOR 13b].

The *closed-loop SI* (CLSI) has a different goal, namely to provide estimated models by taking into consideration the regulator, so as to improve the performance of the global system through redesign of this regulator. Linear identification methods from this category are presented, for instance, in [LAN 93, LAN 97, POP 98, DAU 04, LAN 01] or [LAN 05a].

While nonlinear identification techniques have been proposed in the scientific literature [HAB 99, HUA 04], they are fairly complex, and, as such, they have not been adopted into common industrial applications. Since this book concerns the various applications of Automatic Control in industrial settings, we shall not delve beyond the limits of linear techniques in the modeling of real processes.

Before dealing with CLSI techniques, it is important to explain the main reasons behind this choice. First, the OLSI techniques have been extensively described in [BOR 13a] and [BOR 13b]. We strongly suggest that the reader read these bibliographical references (or other similar ones) before starting this chapter. Second, there are a large number of industrial applications where OL identification models have resulted in inferior regulation performance, compared to those achieved when using CL identification methods (see, for example, [LAN 97] and [LAN 01]). However, the success of CLSI techniques is based on the use of the appropriate numerical algorithms, which are sometimes more complex than those used in OLSI algorithms.

2.2. Framework

Generally speaking, there are some typical situations in which CLSI should be used. Notably, CLSI techniques should be applied when the process to identify:

a) is unstable in the absence of a regulator (recall that most models determined by an OL identification technique are stable);

b) has nominal operating points with a significant drift;

c) does not enable the measurement of the control signal (the command) provided by the regulator (it is a true black box);

d) requires an improvement of the regulator (this is usually designed first using an OL identification model);

e) requires systematic maintenance (usually due to a decrease in regulator performance);

f) has to include fault detection capabilities, and, perhaps, to be fault tolerant as well.

A classical schematic representation of a CL system based on an R (regulation), S (sensitivity) and T (tracking) (RST) type regulator is displayed in Figure 2.1.

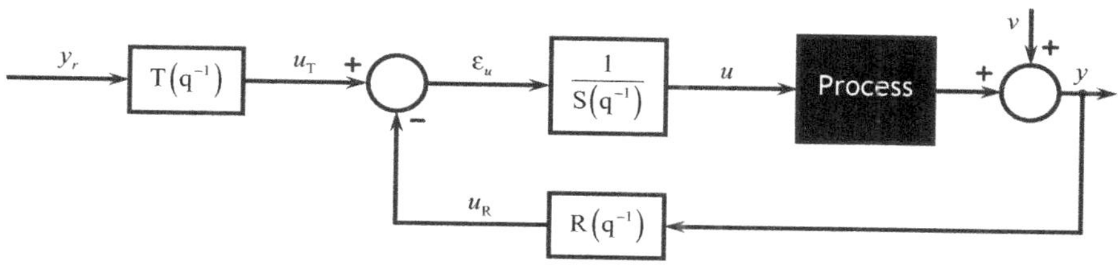

Figure 2.1. *Closed-loop process block diagram based on an RST-type regulator*

The figure has three blocks, not including that of the process:

– the *regulation* block (*a finite impulse response* [FIR] filter):

$$R(q^{-1}) = r_0 + r_1 q^{-1} + r_2 q^{-2} + \cdots + r_{nr} q^{-nr} ; \quad [2.1]$$

– the *sensitivity* adaptation block (an *infinite impulse response* [IIR] filter, which only has poles):

$$\frac{1}{S(q^{-1})} = \frac{1}{s_0 + s_1 q^{-1} + s_2 q^{-2} + \cdots + s_{ns} q^{-ns}} ; \quad [2.2]$$

– the *tracking* block (also an FIR filter):

$$T(q^{-1}) = t_0 + t_1 q^{-1} + t_2 q^{-2} + \cdots + t_{nt} q^{-nt} . \quad [2.3]$$

In definitions [2.1]–[2.3], q^{-1} is the one-step delay operator (one sampling period). The degrees of the three polynomials (nr, ns, nt), as well as their coefficients ($\{r_i\}_{i\in\overline{0,nr}}$, $\{s_j\}_{j\in\overline{0,ns}}$, $\{t_l\}_{l\in\overline{0,nt}}$), can be determined using several algorithms from the literature (see, for example, [LAN 93, POP 98, STE 16]).

However, all of the algorithms rely on a linear process identification model (seen as a black box), expressed by:

$$H(q^{-1}) = \frac{q^{-k}B(q^{-1})}{A(q^{-1})}, \quad [2.4]$$

where $k \in \mathbb{N}^*$ is an approximation of the pure delay of the process (at least equal to one sampling step), while:

$$\begin{cases} B(q^{-1}) = b_0 + b_1 q^{-1} + b_2 q^{-2} + \cdots + b_{nb} q^{-nb} \quad (b_0 \neq 0) \\ A(q^{-1}) = 1 + a_1 q^{-1} + a_2 q^{-2} + \cdots + a_{na} q^{-na} \end{cases} \quad [2.5]$$

are the two polynomials of the associated identification model (with natural notations).

It is easy to prove that, if the process block of Figure 2.1 is replaced by the block of model [2.4], the CL system is then characterized by the following rational *input–output* (I/O) operator (in the absence of exogenous noise v):

$$H_{BF}(q^{-1}) = \frac{q^{-k}B(q^{-1})T(q^{-1})}{A(q^{-1})S(q^{-1}) + q^{-k}B(q^{-1})R(q^{-1})}. \quad [2.6]$$

Expression [2.6] shows how the RST regulator works: it cancels (some or all) the poles of model [2.4] (the roots of polynomial A) and replaces them with other poles, usually chosen as a result of requirements imposed on the CL system (stability, robustness, efficiency, yield, etc.). The desired poles are actually the roots of the polynomial:

$$P(q^{-1}) = A(q^{-1})S(q^{-1}) + q^{-k}B(q^{-1})R(q^{-1}), \quad [2.7]$$

which helps determine polynomials S and R of the regulator. The last polynomial, T, can be determined by imposing the zeros of the CL system, which are given by the polynomial:

$$Q(q^{-1}) = B(q^{-1})T(q^{-1}). \quad [2.8]$$

Usually, the imposed zeros are not only able to change the dynamics of the CL system in the transitory regime, but also the range of the characteristic frequency bandwidth (in position and width).

An I/O operator that characterizes the RST regulator (standing before the process, with the reference y_r as input and the command u as output, in the absence of output noise v) is expressed as follows:

$$\begin{aligned} H^0_{RST}(q^{-1}) &= \frac{A(q^{-1})T(q^{-1})}{A(q^{-1})S(q^{-1})+q^{-k}B(q^{-1})R(q^{-1})} = \frac{A(q^{-1})T(q^{-1})}{P(q^{-1})} \\ &= \frac{Q(q^{-1})}{P(q^{-1})}\cdot\frac{A(q^{-1})}{B(q^{-1})}. \end{aligned} \quad [2.9]$$

This shows how polynomial A is cancelled (it is one of the nominator factors and constitutes the denominator of identification model). However, relation [2.9] cannot be implemented, because it does not take into consideration the real output of the process, but rather the output of the identification model. The real I/O operator is a vector, since the regulator has two inputs: the process output (y) and the CL system reference (y_r):

$$H_{RST}(q^{-1}) = \begin{bmatrix} \dfrac{-R(q^{-1})}{S(q^{-1})} & \dfrac{T(q^{-1})}{S(q^{-1})} \end{bmatrix}. \quad [2.10]$$

Clearly, the control is provided by:

$$u \equiv \begin{bmatrix} \dfrac{-R(q^{-1})}{S(q^{-1})} & \dfrac{T(q^{-1})}{S(q^{-1})} \end{bmatrix} \begin{bmatrix} y \\ y_r \end{bmatrix} \equiv \frac{T(q^{-1})}{S(q^{-1})} y_r - \frac{R(q^{-1})}{S(q^{-1})} y. \quad [2.11]$$

The I/O expression [2.11] of the RST regulator does not change if the disturbance v is taken into account, which leads to the implementation scheme shown in Figure 2.2.

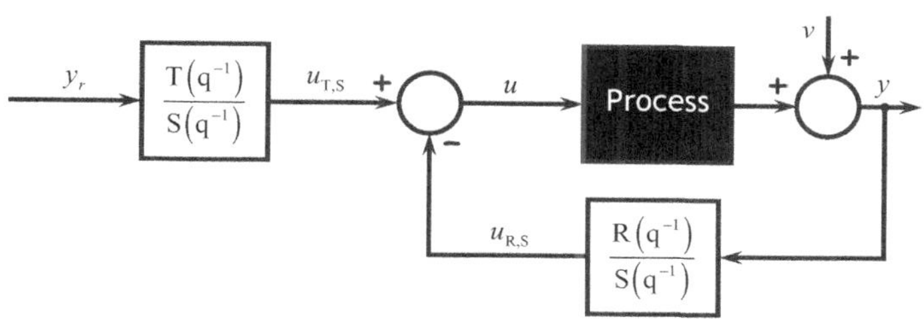

Figure 2.2. *Implementation block diagram based on the RST-type regulator*

With disturbances, the I/O transfer of the CL system is described by:

$$y \equiv \underbrace{\frac{q^{-k}B(q^{-1})T(q^{-1})}{P(q^{-1})}}_{\text{useful filter}} r_y + \underbrace{\frac{A(q^{-1})S(q^{-1})}{P(q^{-1})}}_{\text{disturbances filter}} v. \qquad [2.12]$$

Equation [2.12] shows that the CL model has two filters: one for the useful part (process), which deals with the reference r_y and another for the disturbed part, which deals with the noise v. The two filters have the same poles, as set by the polynomial P (defined in [2.7]). However, their zeros are usually very different. If the reference is a *pseudo-random signal* (PRS) close to white noise and the useful filter has a maximum in the zone of the hodograph (Nyquist plot) close to the critical point (–1,0), then the amplitude of the filtered reference is strengthened in this frequency area. However, as expression [2.11] clearly shows, the control command is affected by the noise v (which is a component of the measured output y). This can introduce significant bias when attempting to identify the process using OL identification techniques.

Two problems can then be formulated in this context: one involving automatic regulation (design the RST regulator) and the other involving the identification of the process (determine a model of the process, especially in CL). The two problems are intercorrelated, since the model of the process determines the regulator, while the regulator in turn determines the model in CL. This complicates the approach used to find solutions. In reality, the succession of these two problems can be cleverly exploited in order to gradually improve the two solutions: the regulator and the model of the process.

More specifically, the CLSI problem can be formulated as follows: starting with a CL process, one seeks to determine a model [2.4] that, with the same regulator, produces a system function as close as possible to the one corresponding to the CL process. A good quality of the identified model is especially needed in the frequency bands that are critical for the control, notably where the hodograph of the transfer function in OL is close to the critical point.

The problem of automatic regulation is formulated as follows: starting with a CL identification model, improve the current regulator (e.g. of RST type), especially around critical points, so as to enforce the process to follow the specified reference as closely as possible, despite the disturbances that affect overall dynamics.

The family of identification methods, collectively referred to as *closed-loop output error* (CLOE), provides a solution to the CLSI problem (which constitutes the central discussion point of this chapter). These methods use output error in a CL configuration in order to come up with appropriate identification methods. They are described in the following section. The quality of the identified model strongly depends on the manner in which the two pieces of information contained in the measured output data (useful and parasite – see once again expression [2.12]) are separated. Since there is no ideal method to carry out this separation, one looks for an optimal or at least satisfactory method (with respect to a criterion that minimizes the output error variance).

2.3. Preliminary identification of a CL process

Before starting the identification process (and especially the updating) of a model in CL with one of the methods from CLOE class, it is useful to provide a first approximation of this model, by using a collection of measured I/O data. To do so, a multivariable approach to the SI can be used. Two cases are analyzed, according to the implementation diagram shown in Figure 2.2:

1) the control signal (the command) u can be measured, as well as the output y;

2) the reference y_r and the output y can be measured, but the control signal u is inaccessible.

The first case is relatively rare. However, if the command is measurable, one of the OLSI methods can be used in order to propose a process model [2.4]. This model presents some obvious disadvantages, the most important of which being its reduced generality (due to the fact that the command is probably not a PRS). Moreover, identifying the model of the process is not enough. The RST regulator has also to be identified in the same way.

The second case is more common in industry. If possible, five polynomials (A, B, R, S and T) are to be determined at the same time by using the measured reference and output.

In both cases, the identification goal can be reached (at least to a satisfying degree) using a multivariable approach. The corresponding identification method belongs to the class of *least squares methods* (LSM).

2.3.1. *Multivariable linear identification methods*

The basic characteristic of models with multiple inputs and outputs (multivariable) is their high complexity. This term can be applied as much to the formalization as to (and especially to) the estimation methodology. In this section, discussion is focused on the *multi-input multi-output* (MIMO) linear model.

One starts with the definition of a *single-input single-output* (SISO) linear model, expressed in linear regression form, as shown, for instance, in [STE 05, BOR 13a] or [BOR 13b]. The description of a MIMO linear model can be generalized as follows:

$$\begin{cases} \mathbf{y}[n] = \boldsymbol{\Theta}\boldsymbol{\varphi}[n] + \mathbf{v}[n] \\ E\{\mathbf{v}[n]\mathbf{v}^T[m]\} = \mathbf{V}\delta_0[n-m] \end{cases}, \quad \forall n, m \in \mathbb{N}, \quad [2.13]$$

where $\mathbf{y} \in \mathbb{R}^{ny}$ is the vector of output signals, $\mathbf{v} \in \mathbb{R}^{ny}$ is the disturbances vector, $\boldsymbol{\Theta} \in \mathbb{R}^{ny \times n\theta}$ is the matrix of unknown parameters (with a structure that is discussed later in this section), $\mathbf{V} \in \mathbb{R}^{ny \times ny}$ is the matrix of disturbances autocovariance (also unknown, but positively defined), $\boldsymbol{\varphi} \in \mathbb{R}^{n\theta}$ is the vector of regressors (which includes measured and possibly estimated data), whereas E is the mathematical expectations operator. The size of the model output vector (*ny*) is regularly smaller than or equal to the number of columns in the parameters matrix ($n\theta$). If the main block of matrix $\boldsymbol{\Theta}$ is diagonal, then the model outputs are uncoupled. In this particular (and rare) case, the MIMO model is broken down into ny independent SISO-type models. If the matrix $\mathbf{V}$ is diagonal, then the disturbances vector only includes a set of uncorrelated white noises.

The ARMAX class [SOD 89, LJU 99, STE 05] is the main candidate for this type of generalization. The equation that corresponds to the MIMO model ARMAX [*na*,*nb*,*nc*] is then written as follows:

$$\begin{aligned} &\mathbf{y}[n] + \mathbf{A}_1\mathbf{y}[n-1] + \cdots + \mathbf{A}_{na}\mathbf{y}[n-na] \\ &\quad = \mathbf{B}_1\mathbf{u}[n-1] + \cdots + \mathbf{B}_{nb}\mathbf{u}[n-nb] \\ &\quad\quad + \mathbf{e}[n] + \mathbf{C}_1\mathbf{e}[n-1] + \cdots + \mathbf{C}_{nc}\mathbf{e}[n-nc], \ \forall n \in \mathbb{N}. \end{aligned} \quad [2.14]$$

In definition [2.14], the scalar parameters (of SISO model) have been replaced by the matrix parameters: $\mathbf{A}_i \in \mathbb{R}^{ny \times ny}$, $\forall i \in \overline{1,na}$; $\mathbf{B}_j \in \mathbb{R}^{ny \times nu}$, $\forall j \in \overline{1,nb}$; $\mathbf{C}_k \in \mathbb{R}^{ny \times ny}$, $\forall k \in \overline{1,nc}$. The number of main parameters is then:

$$n\theta = (na + nc) \cdot ny^2 + nb \cdot nu \cdot ny \ . \quad [2.15]$$

This number has relatively large values even in the case of simple models. To get an idea of its size, let us consider the MIMO-ARMAX[2,2,2] model as an example. It has 24 main parameters, to which the four auxiliary parameters of the noises autocovariance matrix are added.

The model [2.14] is expressed in a more compact manner as:

$$\mathbf{A}\left(q^{-1}\right)\mathbf{y}[n]=\mathbf{B}\left(q^{-1}\right)\mathbf{u}[n]+\mathbf{C}\left(q^{-1}\right)\mathbf{e}[n],\quad \forall n\in\mathbb{N}, \tag{2.16}$$

where $\mathbf{A}\in\mathbb{R}^{ny\times ny}\left(q^{-1}\right)$, $\mathbf{B}\in\mathbb{R}^{ny\times nu}\left(q^{-1}\right)$ and $\mathbf{C}\in\mathbb{R}^{ny\times ny}\left(q^{-1}\right)$ are matrices of polynomials (each element is a polynomial). Note that all the polynomials of a matrix have the same degree (na , nb or nc).

The compact expression [2.16] is useful to find the linear regression form of the MIMO-ARMAX models. For example, a MIMO-ARMAX model with two outputs is expressed using the following equivalent equations (the last one, [2.19], being written in linear regression form):

$$\underbrace{\begin{bmatrix} A_{11}\left(q^{-1}\right) & A_{12}\left(q^{-1}\right) \\ A_{21}\left(q^{-1}\right) & A_{22}\left(q^{-1}\right)\end{bmatrix}}_{\mathbf{A}\left(q^{-1}\right)}\underbrace{\begin{bmatrix} y_1[n] \\ y_2[n]\end{bmatrix}}_{\mathbf{y}[n]}=\underbrace{\begin{bmatrix} C_{11}\left(q^{-1}\right) & C_{12}\left(q^{-1}\right) \\ C_{21}\left(q^{-1}\right) & C_{22}\left(q^{-1}\right)\end{bmatrix}}_{\mathbf{C}\left(q^{-1}\right)}\underbrace{\begin{bmatrix} e_1[n] \\ e_2[n]\end{bmatrix}}_{\mathbf{e}[n]},\quad \forall n\in\mathbb{N}\;; \tag{2.17}$$

$$\begin{bmatrix} y_1[n] \\ y_2[n]\end{bmatrix}+\begin{bmatrix} a_{11,1} & a_{12,1} \\ a_{21,1} & a_{22,1}\end{bmatrix}\begin{bmatrix} y_1[n-1] \\ y_2[n-1]\end{bmatrix}+\cdots+\begin{bmatrix} a_{11,na} & a_{12,na} \\ a_{21,na} & a_{22,na}\end{bmatrix}\begin{bmatrix} y_1[n-na] \\ y_2[n-na]\end{bmatrix}$$
$$=\begin{bmatrix} e_1[n] \\ e_2[n]\end{bmatrix}+\begin{bmatrix} c_{11,1} & c_{12,1} \\ c_{21,1} & c_{22,1}\end{bmatrix}\begin{bmatrix} e_1[n-1] \\ e_2[n-1]\end{bmatrix}+\cdots+\begin{bmatrix} c_{11,nc} & c_{12,nc} \\ c_{21,nc} & c_{22,nc}\end{bmatrix}\begin{bmatrix} e_1[n-nc] \\ e_2[n-nc]\end{bmatrix},\quad \forall n\in\mathbb{N}; \tag{2.18}$$

$$\underbrace{\begin{bmatrix} A_{11}\left(q^{-1}\right) & A_{12}\left(q^{-1}\right) \\ A_{21}\left(q^{-1}\right) & A_{22}\left(q^{-1}\right)\end{bmatrix}}_{\mathbf{A}\left(q^{-1}\right)}\underbrace{\begin{bmatrix} y_1[n] \\ y_2[n]\end{bmatrix}}_{\mathbf{y}[n]}=\underbrace{\begin{bmatrix} C_{11}\left(q^{-1}\right) & C_{12}\left(q^{-1}\right) \\ C_{21}\left(q^{-1}\right) & C_{22}\left(q^{-1}\right)\end{bmatrix}}_{\mathbf{C}\left(q^{-1}\right)}\underbrace{\begin{bmatrix} e_1[n] \\ e_2[n]\end{bmatrix}}_{\mathbf{e}[n]}\begin{bmatrix} y_1[n] \\ y_2[n]\end{bmatrix}$$

$$=\underbrace{\left[\begin{array}{ccc|ccc|ccc|ccc} a_{11,1} & \cdots & a_{11,na} & a_{12,1} & \cdots & a_{12,na} & c_{11,1} & \cdots & c_{11,nc} & c_{12,1} & \cdots & c_{12,nc} \\ \hline a_{21,1} & \cdots & a_{21,na} & a_{22,1} & \cdots & a_{22,na} & c_{21,1} & \cdots & c_{21,nc} & c_{22,1} & \cdots & c_{22,nc}\end{array}\right]}_{\Theta}\underbrace{\begin{bmatrix} -y_1[n-1] \\ \vdots \\ -y_1[n-na] \\ \hline -y_2[n-1] \\ \vdots \\ -y_2[n-na] \\ \hline e_1[n-1] \\ \vdots \\ e_1[n-nc] \\ \hline e_2[n-1] \\ \vdots \\ e_2[n-nc]\end{bmatrix}}_{\boldsymbol{\varphi}[n]}+\begin{bmatrix} e_1[n] \\ e_2[n]\end{bmatrix},\quad \forall n\in\mathbb{N}\,. \tag{2.19}$$

The parameters matrix is non-uniquely determined by equations [2.17] of the model (this is a general characteristic of the MIMO-ARMAX models class). It mainly depends on the configuration of corresponding regressors vector. The considerable size of these variables (parameters matrix and regressors vector) is the main reason for the complexity of the model and the methods employed in its identification.

The stability of such an identification model is also difficult to analyze. The construction of the stability domain $\mathcal{S} \subseteq \mathbb{R}^{ny \times n\theta}$ can no longer benefit from the general results of the SISO models. However, in some simple cases, this domain can be delimited with sufficient accuracy. Obviously, in the case of decoupled models, each I/O channel has its own domain of stability.

In order to make the task of estimating the parameters of the MIMO models easier, an interesting point of view has been adopted, similar to the one that can be found in the MATLAB programming environment. Each output of the MIMO model is seen as a measurement channel that gathers the contributions from all of the inputs. In other words, the global MIMO model is replaced by a collection of *ny* partial *multi-input single-output* (MISO) models. One has to note that the collection of MISO models is not equivalent to the starting model, which was of MIMO type, as each output of the latter can be further influenced by the other outputs. Nevertheless, since each output of the MISO models is determined by all of the inputs, the MATLAB style simplification seems plausible and can be adopted. This approach presents a considerable advantage: it is easier (and more efficient) to identify the *ny* MISO modes (successively or in parallel) at the expense of overall accuracy, than to identify the global MIMO model.

2.3.2. *Estimation of linear MIMO models using the LSM*

The framework defined in the case of SISO models [BOR 13a, BOR 13b] can be adapted to the case of MIMO models. One starts from the equation of the data-providing process:

$$\mathcal{P}(\mathbf{\Theta}^*): \ \mathbf{y}[n] = \mathbf{\Theta}^* \boldsymbol{\varphi}[n] + \mathbf{v}[n], \ \forall n \in \mathbb{N}, \qquad [2.20]$$

with the notations defined in the previous section. As usual, $\mathbf{\Theta}^*$ is the matrix of the true parameters. The disturbances autocovariance matrix is $\mathbf{V}$ (see definition [2.13]). The associated identification model is expressed through a similar equation:

$$\mathcal{M}(\mathbf{\Theta}): \ \mathbf{y}[n] = \mathbf{\Theta} \boldsymbol{\varphi}[n] + \boldsymbol{\varepsilon}[n, \mathbf{\Theta}], \quad \forall n \in \mathbb{N}, \qquad [2.21]$$

where $\boldsymbol{\varepsilon}[n,\boldsymbol{\Theta}] \in \mathbb{R}^{ny}$ is the vector of current modeling errors. If the matrix of unknown parameters is replaced by a matrix of true parameters, then the modeling errors are equal to the noises corrupting the process output:

$$\boldsymbol{\varepsilon}\left[n,\boldsymbol{\Theta}^*\right] = \mathbf{v}[n], \quad \forall n \in \mathbb{N}. \qquad [2.22]$$

In this new framework, the identification problem based on the LSM reduces to the following general and ideal problem of quadratic optimization:

$$\widehat{\boldsymbol{\Theta}} = \underset{\boldsymbol{\Theta}\in\mathcal{S}}{\operatorname{argmin}}\, \mathcal{V}(\boldsymbol{\Theta}) \overset{def}{=} \underset{\boldsymbol{\Theta}\in\mathcal{S}}{\operatorname{argmin}}\, E\left\{\boldsymbol{\varepsilon}[n,\boldsymbol{\Theta}]\boldsymbol{\varepsilon}^T[n,\boldsymbol{\Theta}]\right\}. \qquad [2.23]$$

Unlike the cost function that has to be minimized in the case of SISO models, which is scalar, $\mathcal{V}$ from [2.23] is a matrix cost function. The minimization of a matrix is seemingly unusual. However, following a well-known convention, two matrices can be ordered through the property of positiveness. The square matrix $\mathbf{A}$ is said to be greater than or equal to matrix $\mathbf{B}$ and one writes $\mathbf{A} \geq \mathbf{B}$, if their difference $\mathbf{A}-\mathbf{B}$ is a positively defined matrix: $\mathbf{A}-\mathbf{B} \geq 0$. In the case of matrices from [2.23], the following equivalence can be written:

$$\mathcal{V}(\boldsymbol{\Theta}_1) \leq \mathcal{V}(\boldsymbol{\Theta}_2) \quad \Leftrightarrow \quad \mathcal{V}(\boldsymbol{\Theta}_2) - \mathcal{V}(\boldsymbol{\Theta}_1) \geq 0. \qquad [2.24]$$

Despite this complication, the quadratic expression of the matrix cost function allows finding an elegant solution (albeit a suboptimal one) to problem [2.23], as in the case of SISO models. A natural approach is based on the relaxation of problem [2.23], achieved when replacing the matrix cost function by a related scalar criterion. A simpler optimization problem is then expressed as below:

$$\widehat{\boldsymbol{\Theta}} = \underset{\boldsymbol{\Theta}\in\mathcal{S}}{\operatorname{argmin}}\, \mathcal{V}_0(\boldsymbol{\Theta}) \overset{def}{=} \underset{\boldsymbol{\Theta}\in\mathcal{S}}{\operatorname{argmin}}\, E\left\{\boldsymbol{\varepsilon}^T[n,\boldsymbol{\Theta}]\boldsymbol{\varepsilon}[n,\boldsymbol{\Theta}]\right\} = \underset{\boldsymbol{\Theta}\in\mathcal{S}}{\operatorname{argmin}} \left\|\boldsymbol{\varepsilon}[n,\boldsymbol{\Theta}]\right\|^2. \qquad [2.25]$$

Practically speaking, in [2.25] the external product of the model errors has simply been replaced by their scalar product. Moreover, the price paid for this simplification is not that high. While it is true that problem [2.25] is less concordant with the framework than problem [2.23], it opens up the possibility of finding a solution in a relatively simple manner. Furthermore, there is a link between the two problems. Problem [2.23] is "stronger" than problem [2.25], as minimizing a matrix also leads to the minimization of its *trace* (the sum of the elements on the main diagonal). Obviously, the scalar criterion of problem [2.25] is indeed the trace of the matrix cost function in problem [2.23].

The new definition of cost function eases the search for a solution, which will be an *ideal solution* (but a suboptimal one). Moreover, the ideal solution is deterministic in nature. If the ideal solution is $\mathbf{\Theta}^*$, then the corresponding quadratic cost function has a value equal to disturbances autocovariance matrix:

$$\mathcal{V}\left(\mathbf{\Theta}^*\right) = E\left\{\boldsymbol{\varepsilon}\left[n,\mathbf{\Theta}^*\right]\boldsymbol{\varepsilon}^T\left[n,\mathbf{\Theta}^*\right]\right\} = E\left\{\mathbf{v}[n]\mathbf{v}^T[n]\right\} = \mathbf{V}\,. \qquad [2.26]$$

As a result, once the problem [2.25] is solved, the optimal value of the criterion constitutes an estimation of the disturbances autocovariance matrix. Solving problem [2.25] actually involves determining the (sub)optimal estimations for both the principal and auxiliary parameters.

The practical solution to problem [2.25] (meaning the solution that can be implemented on a computer) is then expressed following the formal expression of the ideal solution. It is of a non-deterministic character (there is a solution for each of the acquired datasets). The following result provides the complete but ideal solution to problem [2.25]. Its proof relies on the generalized method of pseudo-inverses and can be a useful exercise for the reader.

THEOREM 2.1. *The following matrices are defined*:

– $\mathbf{R} \overset{def}{=} E\left\{\boldsymbol{\varphi}[n]\boldsymbol{\varphi}^T[n]\right\} \in \mathbb{R}^{n\theta\times n\theta}$ (*invertible*);

– $\mathbf{S} \overset{def}{=} E\left\{\mathbf{y}[n]\boldsymbol{\varphi}^T[n]\right\} \in \mathbb{R}^{ny\times n\theta}$ (*rectangular*);

– $\mathbf{Y} \overset{def}{=} E\left\{\mathbf{y}[n]\mathbf{y}^T[n]\right\} \in \mathbb{R}^{ny\times ny}$ (*quadratic*).

Then, the ideal solution to the problem of quadratic optimization [2.25] is expressed through the following equations:

$$\mathbf{\Theta}_{\min} = \mathbf{S}\cdot\mathbf{R}^{-1} = \left[E\left\{\mathbf{y}[n]\boldsymbol{\varphi}^T[n]\right\}\right]\left[E\left\{\boldsymbol{\varphi}[n]\boldsymbol{\varphi}^T[n]\right\}\right]^{-1}\ ; \qquad [2.27]$$

$$\mathbf{v}_{\min} = \boldsymbol{\varepsilon}\left[n,\mathbf{\Theta}_{\min}\right] = \mathbf{y}[n] - \mathbf{\Theta}_{\min}\boldsymbol{\varphi}[n],\quad \forall n\in\mathbb{N}\ ; \qquad [2.28]$$

$$\mathcal{V}\left(\mathbf{\Theta}_{\min}\right) = \mathbf{V}_{\min} = \mathbf{Y} - \mathbf{\Theta}_{\min}\mathbf{S}^T = \mathbf{Y} - \mathbf{S}\mathbf{R}^{-1}\mathbf{S}^T \geq 0\,. \qquad [2.29]$$

The conclusion of Theorem 2.1 makes up the core of *multivariable LSM* (MV-LSM). It is very similar to classical LSM for SISO models (see [BOR 13a] or [BOR 13b]). However, it can be noted that the product of the matrices $\mathbf{S}$ and $\mathbf{R}^{-1}$ is carried out in the inverse order for MV-LSM compared to classical LSM (where $\mathbf{R}^{-1}$ appears as the left factor).

The ideal solution [2.27] can be turned into a practical solution due to the ergodic hypothesis [BOR 13a, BOR 13b]. The ideal matrices defined in Theorem 2.1 can be estimated through corresponding temporal means from an acquired dataset, $\mathcal{D}_N = \{\mathbf{y}[n], \boldsymbol{\varphi}[n]\}_{n\in\overline{1,N}}$. More specifically, their estimations are expressed as follows:

$$\left[\begin{array}{l} \hat{\mathbf{R}}_N = \dfrac{1}{N}\sum_{n=1}^{N}\boldsymbol{\varphi}[n]\boldsymbol{\varphi}^T[n] \underset{N\to\infty}{\to} \mathbf{R} \\ \hat{\mathbf{S}}_N = \dfrac{1}{N}\sum_{n=1}^{N}\mathbf{y}[n]\boldsymbol{\varphi}^T[n] \underset{N\to\infty}{\to} \mathbf{S} \\ \hat{\mathbf{Y}}_N = \dfrac{1}{N}\sum_{n=1}^{N}\mathbf{y}[n]\mathbf{y}^T[n] \underset{N\to\infty}{\to} \mathbf{Y} \end{array}\right. . \qquad [2.30]$$

By convention, in SI, the *estimated* models (determined in an approximate manner through an *estimator*) are written with the ^ (hat) sign.

From [2.27]–[2.30], it follows that the implementation (practical) equations of MV-LSM are the following:

$$\hat{\boldsymbol{\Theta}}_N = \hat{\mathbf{S}}_N\hat{\mathbf{R}}_N^{-1} = \left[\frac{1}{N}\sum_{n=1}^{N}\mathbf{y}[n]\boldsymbol{\varphi}^T[n]\right]\left[\frac{1}{N}\sum_{n=1}^{N}\boldsymbol{\varphi}[n]\boldsymbol{\varphi}^T[n]\right]^{-1} ; \qquad [2.31]$$

$$\hat{\mathbf{v}}_N = \boldsymbol{\varepsilon}\left[n, \hat{\boldsymbol{\Theta}}_N\right] = \mathbf{y}[n] - \hat{\boldsymbol{\Theta}}_N\boldsymbol{\varphi}[n] , \quad \forall n \in \mathbb{N} ; \qquad [2.32]$$

$$\hat{\mathbf{V}}_N = \hat{\mathbf{Y}}_N - \hat{\mathbf{S}}_N\hat{\mathbf{R}}_N^{-1}\hat{\mathbf{S}}_N^T . \qquad [2.33]$$

The implementation of equations [2.31]–[2.33] requires the inversion of matrix $\hat{\mathbf{R}}_N$, whose size is determined by the number of parameters to be estimated ($n\theta$). However, the computational effort required for inversion can increase very quickly because this number is relatively large (even for models with a small number of inputs and outputs). As a result, it is desirable that the inversion be performed in a recurrent manner. Generally speaking, the strategy to follow is given by the adaptive LSM [STE 05, BOR 13a, BOR 13b], where Sherman–Morrison's lemma [SHE 50] is the main player. Considering the expression of $\hat{\mathbf{R}}_N$ (see the definitions in [2.30]), this lemma has to be applied successively in order to reduce the computational effort.

The use of the ergodic hypothesis is not enough to ensure the consistency of the SI estimations. For this reason, results that prove the consistency are required. In the case of the MV-LSM, the following theorem holds true.

THEOREM 2.2. *In the context of MV-LSM, the estimations of [2.31]–[2.33] are consistent if the following conditions are met*:

– *the matrices* $\hat{\mathbf{R}}_N$ *are invertible for a large number of acquired data (meaning, if* N *is big enough)*;

– *the exogenous disturbances are ideally uncorrelated with the regressors vector* ($\lim\limits_{N\to\infty}\frac{1}{N}\sum\limits_{n=1}^{N}\mathbf{v}[n]\boldsymbol{\varphi}^T[n]=0$).

While the second condition of Theorem 2.2 is difficult to meet (at first glance), in reality, the regressors vector at the current instant, $\boldsymbol{\varphi}[n]$, includes the values of the signals at previous instants ($n-1$, $n-2$, ...), which are mainly uncorrelated with the current value of the disturbance, $\mathbf{v}[n]$, especially if this disturbance is close to a white noise (non-autocorrelated).

Note that the estimation errors of the main parameters ($\hat{\boldsymbol{\Theta}}_N$) and of the noise ($\hat{\mathbf{v}}_N$) come from a single source: the approximation of the statistical mean by an arithmetic mean. For the auxiliary parameters (the elements of matrix $\hat{\mathbf{V}}_N$), the estimation errors are caused by an additional source: the approximation of matrices $\mathbf{Y}$ and $\mathbf{S}$, according to the definitions of [2.30]. As a result, the matrix $\mathcal{V}\left(\hat{\boldsymbol{\Theta}}_N\right)$ is different from matrix $\hat{\mathbf{V}}_N$ and probably more accurate (as, to calculate it, only the estimation $\hat{\boldsymbol{\Theta}}_N$ is used).

If the number of main parameters goes beyond a few dozen, this form of the MV-LSM can be discarded, as it becomes too slow. Its reduced efficiency is mainly caused by its non-recursive nature. To improve this, one has to go back to the recurrent relations similar to adaptive LSM. The reader is invited to rediscover these relations for MV-LSM, using a similar reasoning to that leading to the algorithm of adaptive LSM (see [STE 05, BOR 13a, BOR 13b]). The only elements to look out for are the starting point (which is now fixed by the estimation of [2.31]) and the order of the multiplication between matrices. The efficiency of the resulting adaptive algorithm is guaranteed by the same Sherman–Morrison's lemma, which does not need to be generalized at this time.

2.3.3. *Identifying CL processes using the MV-LSM*

The previous sections have prepared the identification techniques described in the following. However, it can be outlined that MV-LSM is employed in other industrial applications as well, e.g. those concerning the OL identification of complex energy production systems, the prediction of intercorrelated meteorological or ecological datasets, etc.

Come back once again to Figure 2.2, and analyze the two cases mentioned earlier. In order to identify the process, as well as the RST regulator, the problem has to be formulated in a way that yields the use of the MV-LSM.

2.3.3.1. *I/O data measured directly from the process*

Since the signals u and y are available to the user, they can be grouped into an output vector $\mathbf{y} \equiv [u \quad y]^T$. The dynamics of the multivariable system are then described by the following equations:

$$\underbrace{\begin{bmatrix} \mathrm{S}(\mathrm{q}^{-1}) & \mathrm{R}(\mathrm{q}^{-1}) \\ -\mathrm{q}^{-k}\mathrm{B}(\mathrm{q}^{-1}) & \mathrm{A}(\mathrm{q}^{-1}) \end{bmatrix}}_{\mathbf{A}(\mathrm{q}^{-1})} \underbrace{\begin{bmatrix} u \\ y \end{bmatrix}}_{\mathbf{y}} \equiv \underbrace{\begin{bmatrix} \mathrm{T}(\mathrm{q}^{-1})y_r \\ \mathrm{A}(\mathrm{q}^{-1})v \end{bmatrix}}_{\mathbf{w}}. \qquad [2.34]$$

This actually expresses an AR multivariable model (a particular case of MIMO-ARMAX model). In order to simplify the approach, one assumes that the polynomials of matrix $\mathbf{A}(\mathrm{q}^{-1})$ have the same degree, *na* (after identification, one can remove the negligible amplitude coefficients that correspond to the monomials of power close to $n\alpha$). Equation [2.34] is then expressed in an equivalent manner, as follows:

$$\mathbf{A}_0\mathbf{y}[n] + \mathbf{A}_1\mathbf{y}[n-1] + \cdots + \mathbf{A}_{n\alpha}\mathbf{y}[n-n\alpha] = \mathbf{w}[n], \quad \forall n \in \mathbb{N}, \qquad [2.35]$$

where $\mathbf{A}_0, \mathbf{A}_1, \ldots, \mathbf{A}_{n\alpha} \in \mathbb{R}^{2\times 2}$ are the unknown parameters of the model (matrices with four elements each, made up of the coefficients taken from the polynomials A, B, R and S). Note that:

$$\mathbf{A}_0 = \begin{bmatrix} s_0 & r_0 \\ 0 & 1 \end{bmatrix}, \qquad [2.36]$$

as the pure delay k is at least equal to 1 (the free coefficient of the polynomial $\mathrm{q}^{-k}\mathrm{B}$ is equal to zero). It follows that the matrix $\mathbf{A}_0$ is invertible (since the coefficient s_0 is non-null). This results in the ability to normalize equation [2.36] by multiplication with the inverse of $\mathbf{A}_0$:

$$\mathbf{y}[n] = -\mathbf{A}_0^{-1}\mathbf{A}_1\mathbf{y}[n-1] - \cdots - \mathbf{A}_0^{-1}\mathbf{A}_{n\alpha}\mathbf{y}[n-n\alpha] + \mathbf{A}_0^{-1}\mathbf{w}[n], \quad \forall n \in \mathbb{N}. \qquad [2.37]$$

One can rewrite the matrix coefficients of equation [2.37] as $\mathbf{A}_1$, $\mathbf{A}_2$, …, $\mathbf{A}_{n\alpha}$ and the noise as $\mathbf{w}$, such that:

$$\mathbf{y}[n] = -\mathbf{A}_1\mathbf{y}[n-1] - \cdots - \mathbf{A}_{n\alpha}\mathbf{y}[n-n\alpha] + \mathbf{w}[n], \quad \forall n \in \mathbb{N}. \qquad [2.38]$$

This latter expression helps to obtain the linear regression form of the model (with a natural notation for the coefficients of polynomial $\mathrm{q}^{-k}\mathrm{B}$):

$$\begin{bmatrix} u[n] \\ y[n] \end{bmatrix} = \underbrace{\left[\begin{array}{ccc|ccc} s_1 & \cdots & s_{n\alpha} & r_1 & \cdots & r_{n\alpha} \\ \hline -b_1^k & \cdots & -b_{n\alpha}^k & a_1 & \cdots & a_{n\alpha} \end{array}\right]}_{\mathbf{\Theta}} \underbrace{\begin{bmatrix} -u[n-1] \\ \vdots \\ -u[n-n\alpha] \\ \hline -y[n-1] \\ \vdots \\ -y[n-n\alpha] \end{bmatrix}}_{\boldsymbol{\varphi}[n]} + \begin{bmatrix} w_1[n] \\ w_2[n] \end{bmatrix}, \quad \forall n \in \mathbb{N}. \qquad [2.39]$$

In the case of model [2.39], the solution [2.31] of the MV-LSM yields direct recovery of the estimated polynomials $\hat{\mathrm{A}}$, $\hat{\mathrm{B}}$, $\hat{\mathrm{R}}$, $\hat{\mathrm{S}}$ (the last two being normalized) and the delay k. After this operation, one can remove the negligible coefficients in four polynomials, in order to estimate their real degrees (this operation being referred to as *dimensional reduction*). Moreover, equation [2.32] helps estimate the noise vector $\mathbf{w}$. From equation [2.34] (to which the inverse of the matrix

$\mathbf{A}_0$ is applied [2.36]), the exogenous disturbance can easily be estimated using a recursive relation:

$$\hat{\mathrm{A}}\left(\mathrm{q}^{-1}\right)\hat{v} \equiv \hat{w}_2 \;\Leftrightarrow\; \hat{v}[n] = -\hat{a}_1\hat{v}[n-1] - \cdots - \hat{a}_{na}\hat{v}[n-na] + \hat{w}_2[n],\ \forall n \in \mathbb{N}. \quad [2.40]$$

The same equation [2.34] leads to:

$$\hat{\mathrm{T}}\left(\mathrm{q}^{-1}\right)y_r \equiv s_0\hat{w}_1 + r_0\hat{w}_2, \quad [2.41]$$

which shows that, to determine the last polynomial of the CL model, T, the reference y_r, as well as the parameters s_0 and r_0, have to be known *a priori*. Normally, the reference is well known in advance (or it can at least be measured), while $s_0 = 1$ and $r_0 = 0$. In these conditions, applying the classic LSM onto equation [2.41] leads to:

$$\begin{bmatrix} \hat{t}_0 \\ \hat{t}_1 \\ \vdots \\ \hat{t}_{nt} \end{bmatrix} = \left(\frac{1}{N}\sum_{n=1}^{N} \begin{bmatrix} y_r[n] \\ y_r[n-1] \\ \vdots \\ y_r[n-nt] \end{bmatrix} \begin{bmatrix} y_r[n] & y_r[n-1] & \cdots & y_r[n-nt] \end{bmatrix} \right)^{-1} \left(\frac{1}{N}\sum_{n=1}^{N} \begin{bmatrix} y_r[n] \\ y_r[n-1] \\ \vdots \\ y_r[n-nt] \end{bmatrix} \hat{w}_1[n] \right). \quad [2.42]$$

The accuracy of the identified model is given by equation [2.33]. The smaller the eigenvalues of matrix $\hat{\mathbf{V}}_N$ (recall that the matrix is positively defined), the more accurate the model.

2.3.3.2. *I/O data measured from the CL system (without access to the process control signal)*

In this case, since the data from signals y_r and y are available, the output vector $\mathbf{y} \equiv [y_r \quad y]^T$ can naturally be built. This results in the following dynamic equation:

$$\underbrace{\begin{bmatrix} \mathrm{T}\left(\mathrm{q}^{-1}\right) & -\mathrm{R}\left(\mathrm{q}^{-1}\right) \\ -\mathrm{q}^{-k}\mathrm{Q}\left(\mathrm{q}^{-1}\right) & \mathrm{P}\left(\mathrm{q}^{-1}\right) \end{bmatrix}}_{\mathbf{A}\left(\mathrm{q}^{-1}\right)} \underbrace{\begin{bmatrix} y_r \\ y \end{bmatrix}}_{\mathbf{y}} \equiv \underbrace{\begin{bmatrix} \mathrm{S}\left(\mathrm{q}^{-1}\right)u \\ \mathrm{A}\left(\mathrm{q}^{-1}\right)\mathrm{S}\left(\mathrm{q}^{-1}\right)v \end{bmatrix}}_{\mathbf{w}}, \quad [2.43]$$

with definitions [2.8] and [2.7].

As in the previous case, the linear regression form associated with model [2.43] is:

$$\begin{bmatrix} y_r[n] \\ y[n] \end{bmatrix} = \underbrace{\left[\begin{array}{ccc|ccc} t_1 & \cdots & t_{n\alpha} & -r_1 & \cdots & -r_{n\alpha} \\ \hline -q_1^k & \cdots & -q_{n\alpha}^k & p_1 & \cdots & p_{n\alpha} \end{array}\right]}_{\mathbf{\Theta}} \underbrace{\begin{bmatrix} -y_r[n-1] \\ \vdots \\ -y_r[n-n\alpha] \\ \hline -y[n-1] \\ \vdots \\ -y[n-n\alpha] \end{bmatrix}}_{\boldsymbol{\varphi}[n]} + \begin{bmatrix} w_1[n] \\ w_2[n] \end{bmatrix}, \forall n \in \mathbb{N} \quad [2.44]$$

with $t_0 = s_0 = 1$ and $r_0 = 0$. The MV-LSM applied to equation [2.44] leads to a direct retrieval of the estimations $\hat{\mathrm{P}}$, $\hat{\mathrm{Q}}$, $\hat{\mathrm{R}}$, $\hat{\mathrm{T}}$ and of the pure delay k. An estimation of the virtual disturbance $\hat{\mathbf{w}}$ is also available.

After performing the dimensional reduction (which involves the removal of the negligible coefficients), the polynomial $\hat{\mathrm{B}}$ is obtained from the following identity (which is actually a Bézout's identity or a Diophantine equation):

$$\hat{\mathrm{Q}}\left(\mathrm{q}^{-1}\right) \equiv \hat{\mathrm{B}}\left(\mathrm{q}^{-1}\right)\hat{\mathrm{T}}\left(\mathrm{q}^{-1}\right), \quad [2.45]$$

where the estimation $\hat{\mathrm{Q}}$ is already available. Next, the product $\widehat{\mathrm{AS}}$ can be estimated straightforwardly:

$$\widehat{\mathrm{AS}}\left(\mathrm{q}^{-1}\right) \equiv \hat{\mathrm{P}}\left(\mathrm{q}^{-1}\right) - \mathrm{q}^{-k}\hat{\mathrm{B}}\left(\mathrm{q}^{-1}\right)\hat{\mathrm{R}}\left(\mathrm{q}^{-1}\right). \quad [2.46]$$

Note that the first coefficient of the polynomial $\widehat{\mathrm{AS}}$ is $\hat{s}_0$ (see definitions [2.2] and [2.5]).

Then, the second element of the right term in equation [2.45] allows estimating the exogenous noise, using a recursive relation (with natural notations):

$$\widehat{\mathrm{AS}}\left(\mathrm{q}^{-1}\right)\hat{v} \equiv \hat{w}_2$$

$$\Leftrightarrow \hat{v}[n] = (\widehat{as})_1 \hat{v}[n-1] - \cdots - (\widehat{as})_{na+ns} \hat{v}[n-na-ns] + \hat{w}_2[n], \; \forall n \in \mathbb{N}. \quad [2.47]$$

The only problem that remains to be solved is the separation of the polynomials $\hat{\mathrm{A}}$ and $\hat{\mathrm{S}}$ from product $\widehat{\mathrm{AS}}$. The first element of the right term in equation [2.43] could be used for this aim as long as the control signal is accessible. However, this is not possible. The only solution that can be considered is to determine the roots of product $\widehat{\mathrm{AS}}$ and to distribute them between the two polynomials empirically. Since in any case $\hat{\mathrm{S}}$ cannot be unstable (otherwise the RTS regulator would become unstable), all the unstable roots of product $\widehat{\mathrm{AS}}$ (meaning with amplitudes equal to 1 or greater) are given to polynomial $\hat{\mathrm{A}}$. Similarly, the degree of polynomial $\hat{\mathrm{A}}$ should be at least equal to or greater than nb (in order to enable the physical realization of the identification model in real time, if required). One assumes that the RST regulator has a double role: to stabilize the process and to improve the robustness of the CL system. As a result, by excluding unstable roots the polynomial $\hat{\mathrm{A}}$ can receive the roots that are the closest to instability in the collection. Consequently, two polynomials $\hat{\mathrm{A}}$ and $\hat{\mathrm{S}}$ can be proposed, even though some uncertainty regarding their degrees and coefficients remains.

Once the polynomial $\hat{\mathrm{S}}$ has been determined, it is possible to estimate the control using a second recursive relation:

$$\hat{\mathrm{S}}\left(\mathrm{q}^{-1}\right)\hat{u} \equiv \hat{w}_1$$

$$\Leftrightarrow \quad \hat{u}[n] = \hat{s}_1\hat{u}[n-1] - \cdots - \hat{s}_{ns}\hat{u}[n-ns] + \hat{w}_1[n], \ \forall n \in \mathbb{N}. \qquad [2.48]$$

Usually, five polynomials estimated using the MV-LSM are not too accurate, but are rather suboptimal, because of the cost function $\mathcal{V}_0$ from problem [2.25], which has been minimized, in order for them to be estimated. They can very well be used as an initialization in an online adaptation algorithm based on a method belonging to the CLOE class.

2.4. CLOE class of identification methods

2.4.1. *Principle of CLOE methods*

The name "CLOE" for the identification methods in this chapter comes from the image in Figure 2.3.

After being identified offline (as presented before), the process model is updated online (at each sampling instant) as a function of the *output error*, denoted by ε_y in the figure. Updating concerns the coefficients of polynomials [2.5]. Sometimes (but seldom), other parameters of model [2.4] can be changed (e.g. the structural indices k, na and/or nb). Clearly, each modification made to the model of process results in a renewal of the RST regulator (even though this does not appear to be the case in the figure).

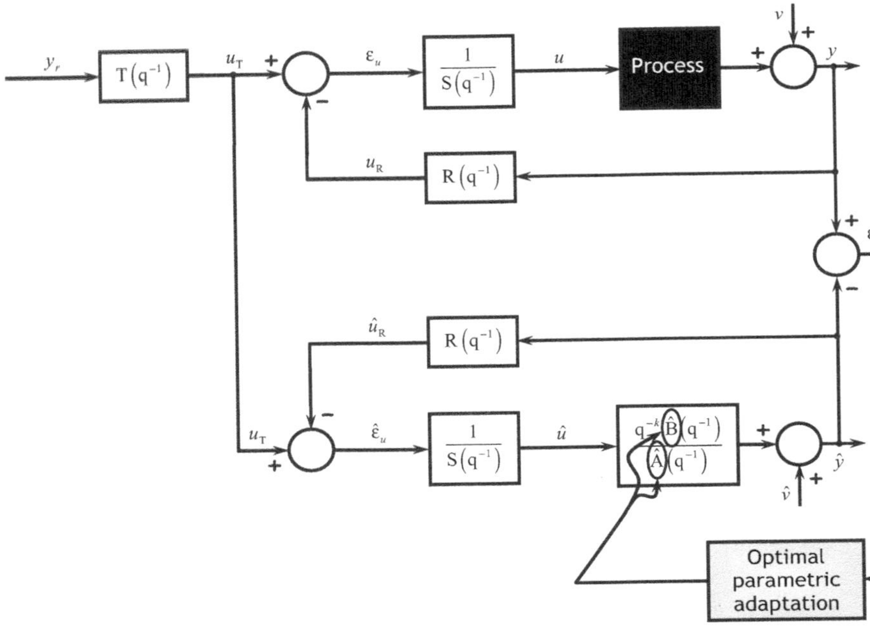

Figure 2.3. *Principle of CLOE identification methods*

The output error in a closed loop actually measures the gap between the real process and its model. The variance of this error has to be minimized, in order to update the identification model.

The main objective of the OLSI is, therefore, to find an optimal model of the process by minimizing the gap between the simulated output and the real output of the process in a CL.

For this type of identification method, it is important to run the identification model in parallel with the process, as both are controlled by the same regulator (RST in this approach) and have the goal of following the same reference, while remaining stable.

In this section, one assumes that the RST is known, while the model of process has to be updated, depending on the current output error.

2.4.2. *Basic CLOE method*

Before presenting the equations of the CLOE method, recall the adaptive (online) least squares algorithm for SISO systems (see, for example, [SCS 05, BOR 13a] or [BOR 13b]). The basic CLOE procedure is strongly based on this algorithm. The framework of this algorithm relies on the SISO linear regression model below:

$$y_{\mathcal{M}}[n,\boldsymbol{\theta}] = \boldsymbol{\varphi}^T[n]\boldsymbol{\theta}, \quad \forall n \in \mathbb{N}, \qquad [2.49]$$

where: $\boldsymbol{\theta} \in \mathbb{R}^{n\theta}$ is the vector of the unknown parameters and $\boldsymbol{\varphi}[n] \in \mathbb{R}^{n\theta}$ is the regressors vector at current instant $n \in \mathbb{N}$. The process is virtually described through an equation similar to [2.49]:

$$y[n] = \boldsymbol{\varphi}^T[n]\boldsymbol{\theta}^* + v[n], \quad \forall n \in \mathbb{N}, \qquad [2.50]$$

where $\boldsymbol{\theta}^* \in \mathbb{R}^{n\theta}$ is the vector of the real parameters (unknown) and v is an exogenous disturbance of dispersion λ_v^2 (also unknown).

The quadratic criterion used to determine an optimal model from a measured (and/or estimated) dataset $\mathcal{D}_N = \{y[n], \boldsymbol{\varphi}[n]\}_{n \in \overline{1,N}}$ is defined as follows:

$$\mathcal{V}_N(\boldsymbol{\theta}) = \sum_{n=1}^{N} \left(y[n] - y_{\mathcal{M}}[n,\boldsymbol{\theta}]\right)^2 = \sum_{n=1}^{N} \left(y[n] - \boldsymbol{\varphi}^T[n]\boldsymbol{\theta}\right)^2. \qquad [2.51]$$

The solution provided by the offline LSM (which minimizes criterion [2.51]) is expressed by the following equations:

$$\hat{\boldsymbol{\theta}}_N = \left(\frac{1}{N}\sum_{n=1}^{N} \boldsymbol{\varphi}[n]\boldsymbol{\varphi}^T[n]\right)^{-1} \left(\frac{1}{N}\sum_{n=1}^{N} \boldsymbol{\varphi}[n]y[n]\right); \qquad [2.52]$$

$$\hat{v}_N[n] = y[n] - \boldsymbol{\varphi}^T[n]\hat{\boldsymbol{\theta}}_N, \quad \forall n \in \mathbb{N}; \qquad [2.53]$$

$$\hat{\lambda}_{v,N}^2 = \frac{1}{N - n\theta}\mathcal{V}_N\left(\hat{\boldsymbol{\theta}}\right) = \frac{1}{N - n\theta}\sum_{n=1}^{N} \hat{v}_N^2 = \frac{1}{N - n\theta}\sum_{n=1}^{N} \left(y[n] - \boldsymbol{\varphi}^T[n]\hat{\boldsymbol{\theta}}\right)^2. \qquad [2.54]$$

The numerical procedure described in the following Algorithm 2.1 starts from expression [2.52] of the solution provided by the offline LSM and uses a clever notation:

$$\mathbf{P}_N = \left(\sum_{n=1}^{N} \boldsymbol{\varphi}[n]\boldsymbol{\varphi}^T[n]\right)^{-1}. \qquad [2.55]$$

1) *Initialization*:

a) Set the structural index of the model: $n\theta \in \mathbb{N}^*$.

b) If possible, acquire a reduced I/O dataset from the black box, namely: $\mathcal{D}_{N_0} = \{(y[n], \boldsymbol{\varphi}[n])\}_{n\in\overline{1,N_0}}$.

c) If the dataset $\mathcal{D}_{N_0}$ is available, use an optimal offline identification method (e.g. the LSM) to estimate the matrix $\mathbf{P}_0$ and the vector $\hat{\boldsymbol{\theta}}_0$.

d) Otherwise, set $\mathbf{P}_0 = \alpha^2 \mathbf{I}_{n\theta}$ and $\hat{\boldsymbol{\theta}}_0$ arbitrarily, but with a large enough value of α (a few hundreds, at least).

e) Set the starting iteration index: $n = 1$.

2) *For* $n \geq 1$:

2.1. Acquire the current I/O data: $\{y[n], \boldsymbol{\varphi}[n]\}$.

2.2. Evaluate the simulation error: $\varepsilon[n] = y[n] - \boldsymbol{\varphi}^T[n]\hat{\boldsymbol{\theta}}_{n-1}$.

2.3. Evaluate the auxiliary vector: $\boldsymbol{\xi}_n = \mathbf{P}_{n-1}\boldsymbol{\varphi}[n]$.

2.4. Evaluate the sensitivity gain: $\boldsymbol{\gamma}_n = \dfrac{\boldsymbol{\xi}_n}{1 + \boldsymbol{\varphi}^T[n]\boldsymbol{\xi}_n}$.

2.5. Update the matrix $\mathbf{P}_{n-1}$, with: $\mathbf{P}_n = \mathbf{P}_{n-1} - \boldsymbol{\gamma}_n \boldsymbol{\xi}_n^T$ (while avoiding the explicit inversion of matrices, using Sherman–Morrison's lemma).

2.6. Update the parameter vector: $\hat{\boldsymbol{\theta}}_n = \hat{\boldsymbol{\theta}}_{n-1} + \boldsymbol{\gamma}_n \varepsilon[n]$.

2.7. Move on to the next iteration: $n \leftarrow n + 1$.

3) *Return the parameters vector* $\hat{\boldsymbol{\theta}}_n$ *with updated values at each instant* $n \geq 0$.

Algorithm 2.1. *Steps of the basic adaptive LSM*

Steps 2.1, 2.4 and 2.5 of Algorithm 2.1 are the milestones of numerical procedure. In the case of model [2.4] (of ARX type, actually), the current regressors vector is expressed as follows:

$$\boldsymbol{\varphi}[n] = [-y[n-1] \ \cdots \ -y[n-na] \ \vdots \ u[n-k] \ \cdots \ u[n-nb-k]]^T. \qquad [2.56]$$

Normally, the output y of the process can be measured. However, the command u is inaccessible. As a result, the control has to be evaluated using the previous RST regulators. More specifically, in the second part of vector [2.56], the regressed values of the command are calculated using the expression below:

$$\hat{u}[n] = \frac{T_{n-1}(q^{-1})}{S_{n-1}(q^{-1})} y_r[n] - \frac{R_{n-1}(q^{-1})}{S_{n-1}(q^{-1})} \hat{y}[n] = \frac{T_{n-1}(q^{-1})}{S_{n-1}(q^{-1})} y_r[n] - \frac{R_{n-1}(q^{-1})}{S_{n-1}(q^{-1})} \hat{\boldsymbol{\varphi}}^T[n]\hat{\boldsymbol{\theta}}_{n-1}, \quad \forall n \in \mathbb{N}, \qquad [2.57]$$

where, by definition:

$$\hat{\boldsymbol{\varphi}}[n] = [-y[n-1] \quad \cdots \quad -y[n-na] \mid \hat{u}[n-k] \quad \cdots \quad \hat{u}[n-nb-k]]^T. \qquad [2.58]$$

Equations [2.57] and [2.58] show that the command evaluation is carried out recursively. The regressors vector is then easily estimated. Note that the reference y_r is measurable.

Since the vector of the unknown parameters coming from model [2.4], denoted by $\hat{\boldsymbol{\theta}}_n$, is variable in time, the CLOE method uses two types of simulated output data:

$$\hat{y}_b[n] = \hat{\boldsymbol{\varphi}}^T[n]\hat{\boldsymbol{\theta}}_{n-1}, \quad \forall n \in \mathbb{N} \text{ (\textit{a priori}, backward);} \qquad [2.59]$$

$$\hat{y}_f[n] = \hat{\boldsymbol{\varphi}}^T[n]\hat{\boldsymbol{\theta}}_n, \quad \forall n \in \mathbb{N} \text{ (\textit{a posteriori}, forward).} \qquad [2.60]$$

In the expressions above, the regressors vector only includes *regressed* I/O values (i.e. evaluated up to the previous instant, but not at the current instant – see again definition [2.56]). In the *a priori* output [2.59], this vector and the vector of the variable parameters are synchronized at the previous instant ($n-1$). On the contrary, in the *a posteriori* output [2.60], the parameters vector is ahead of the regressors vector by one sampling step.

Two types of output error are, therefore, available at each instant:

$$\varepsilon_y^b[n] = y[n] - \hat{y}_b[n] = y[n] - \hat{\boldsymbol{\varphi}}^T[n]\hat{\boldsymbol{\theta}}_{n-1}, \quad \forall n \in \mathbb{N} \text{ (\textit{a priori});} \qquad [2.61]$$

$$\varepsilon_y^f[n] = y[n] - \hat{y}_f[n] = y[n] - \hat{\boldsymbol{\varphi}}^T[n]\hat{\boldsymbol{\theta}}_n, \quad \forall n \in \mathbb{N} \text{ (\textit{a posteriori}).} \qquad [2.62]$$

From Algorithm 2.1, one can deduce that the expression of parameters adaptation is written as follows, at length (with definition [2.61]):

$$
\begin{aligned}
\hat{\boldsymbol{\theta}}_n &= \hat{\boldsymbol{\theta}}_{n-1} + \hat{\boldsymbol{\gamma}}_n \varepsilon_y^b[n] = \hat{\boldsymbol{\theta}}_{n-1} + \frac{\hat{\boldsymbol{\xi}}_n}{1+\hat{\boldsymbol{\varphi}}^T[n]\hat{\boldsymbol{\xi}}_n}\varepsilon_y^b[n] \\
&= \hat{\boldsymbol{\theta}}_{n-1} + \frac{\mathbf{P}_{n-1}\hat{\boldsymbol{\varphi}}[n]}{1+\hat{\boldsymbol{\varphi}}^T[n]\mathbf{P}_{n-1}\hat{\boldsymbol{\varphi}}[n]}\varepsilon_y^b[n], \ \forall n \geq 0.
\end{aligned}
\qquad [2.63]
$$

Expression [2.63] allows deriving a relationship between the two types of output error:

$$
\begin{aligned}
\varepsilon_y^f[n] &= y[n] - \hat{\boldsymbol{\varphi}}^T[n]\hat{\boldsymbol{\theta}}_n \\
&= \underbrace{y[n] - \hat{\boldsymbol{\varphi}}^T[n]\hat{\boldsymbol{\theta}}_{n-1}}_{\varepsilon_y^b[n]} - \frac{\hat{\boldsymbol{\varphi}}^T[n]\mathbf{P}_{n-1}\hat{\boldsymbol{\varphi}}[n]}{1+\hat{\boldsymbol{\varphi}}^T[n]\mathbf{P}_{n-1}\hat{\boldsymbol{\varphi}}[n]}\varepsilon_y^b[n] \\
&= \frac{1+\cancel{\hat{\boldsymbol{\varphi}}^T[n]\mathbf{P}_{n-1}\hat{\boldsymbol{\varphi}}[n]} - \cancel{\hat{\boldsymbol{\varphi}}^T[n]\mathbf{P}_{n-1}\hat{\boldsymbol{\varphi}}[n]}}{1+\hat{\boldsymbol{\varphi}}^T[n]\mathbf{P}_{n-1}\hat{\boldsymbol{\varphi}}[n]}\varepsilon_y^b[n] \\
&= \frac{\varepsilon_y^b[n]}{1+\hat{\boldsymbol{\varphi}}^T[n]\mathbf{P}_{n-1}\hat{\boldsymbol{\varphi}}[n]}, \quad \forall n \geq 0.
\end{aligned}
\qquad [2.64]
$$

It is therefore possible to know the *a posteriori* output error before updating the parameters vector. Moreover, since the matrices $\mathbf{P}_n$ are positively defined, the final result of [2.64] shows that the output error tends to decrease in amplitude (as $1+\hat{\boldsymbol{\varphi}}^T[n]\mathbf{P}_{n-1}\hat{\boldsymbol{\varphi}}[n] > 1$). According to equation [2.64], the main updating relation is expressed as follows:

$$
\hat{\boldsymbol{\theta}}_n = \hat{\boldsymbol{\theta}}_{n-1} + \frac{\mathbf{P}_{n-1}\hat{\boldsymbol{\varphi}}[n]}{1+\hat{\boldsymbol{\varphi}}^T[n]\mathbf{P}_{n-1}\hat{\boldsymbol{\varphi}}[n]}\varepsilon_y^b[n] = \hat{\boldsymbol{\theta}}_{n-1} + \mathbf{P}_{n-1}\hat{\boldsymbol{\varphi}}[n]\varepsilon_y^f[n], \quad \forall n \geq 0. \qquad [2.65]
$$

The *a posteriori* error plays an important role in the efficient implementation of algorithms belonging to the CLOE family.

The update of parameters corresponding to ARX model is therefore carried out following equations [2.65], where the *a posteriori* error depends on the *a priori* error, which can easily be calculated using the parameters vector $\hat{\boldsymbol{\theta}}_{n-1}$ (see definition [2.61]). The algorithm of optimal parameters adaptation proposed by the basic CLOE method is similar to the one used in the online LSM. Its steps are presented in Algorithm 2.2.

1) *Initialization*:

a) Set the structural index of the model: $n\theta = na + nb \in \mathbb{N}^*$.

b) Identify a CL model of the process using one of the offline techniques suggested in the previous section. As a result, a first estimation of five polynomials is available: $\hat{A}_0$, $\hat{B}_0$, $\hat{R}_0$, $\hat{S}_0$ and $\hat{T}_0$. Estimations of the process pure delay (k) and of the command ($\hat{u}$) are also available. This allows building the initial vector of estimated parameters, $\hat{\boldsymbol{\theta}}_0$, by using the coefficients of polynomials $\hat{A}_0$ and $\hat{B}_0$.

c) Choose an initialization for the main matrices: $\mathbf{P}_0 = \alpha^2 \mathbf{I}_{n\theta}$, with a large enough value of α (a few hundreds, at least).

d) Initialize the estimated regressors vector:

$$\hat{\varphi}[1] = \left[-y[N-1] \quad \cdots \quad -y[N-na] \;\vdots\; \hat{u}[N-k] \quad \cdots \quad \hat{u}[N-nb-k]\right]^T,$$

where N is the size of the offline measurement horizon.

e) Set the starting iteration index: $n = 1$.

2) *For* $n \geq 1$:

2.1. Acquire the current reference and output: $\{y_r[n], y[n]\}$.

2.2. Estimate the current command: $\hat{u}[n] = \dfrac{\hat{T}_{n-1}(q^{-1})}{\hat{S}_{n-1}(q^{-1})} y_r[n] - \dfrac{\hat{R}_{n-1}(q^{-1})}{\hat{S}_{n-1}(q^{-1})} \hat{\varphi}^T[n]\hat{\boldsymbol{\theta}}_{n-1}$.

2.3. Prepare the regressors vector for the next iteration:

$$\hat{\varphi}[n+1] = \left[-y[n] \quad \cdots \quad -y[n-na+1] \;\vdots\; \hat{u}[n-k+1] \quad \cdots \quad \hat{u}[n-nb-k+1]\right]^T.$$

(Take values from the vector $\hat{\varphi}[1]$, if necessary.)

2.4. Evaluate the *a priori* output error: $\varepsilon_y^b[n] = y[n] - \hat{\varphi}^T[n]\hat{\boldsymbol{\theta}}_{n-1}$.

2.5. Evaluate the auxiliary vector: $\hat{\xi}_n = \mathbf{P}_{n-1}\hat{\varphi}[n]$.

2.6. Evaluate the multiplicity constant of the output errors: $\hat{\mu}_n = \dfrac{1}{1 + \hat{\varphi}^T[n]\hat{\xi}_n}$ (scalar).

2.7. Evaluate the *a posteriori* output error: $\varepsilon_y^f[n] = \hat{\mu}_n \varepsilon_y^b[n]$.

2.8. Update the main matrix: $\mathbf{P}_n = \mathbf{P}_{n-1} - \hat{\mu}_n \hat{\xi}_n \hat{\xi}_n^T$.

2.9. Update the parameters vector: $\hat{\boldsymbol{\theta}}_n = \hat{\boldsymbol{\theta}}_{n-1} + \hat{\xi}_n \varepsilon_y^f[n]$.

2.10. Update the RST regulator: $\left\{\hat{R}_n, \hat{S}_n, \hat{T}_n\right\}$ from the model $\left\{\hat{A}_n, \hat{B}_n\right\}$ (built from $\hat{\boldsymbol{\theta}}_n$).

2.11. Move on to the next iteration: $n \leftarrow n+1$.

3) *Return the model of CL process* $\left\{\hat{A}_n, \hat{B}_n, \hat{R}_n, \hat{S}_n, \hat{T}_n\right\}$ *at each instant* $n \geq 0$. *An estimation of the command* $\hat{u}$ *is also available.*

Algorithm 2.2. *Steps of the basic CLOE procedure*

Updating the RST regulator (required in step 2.10 of Algorithm 2.2) is carried out as detailed in Chapters 3 and 4 of this book, for example. Overall, this procedure is quite effective. Nevertheless, the reader can see some possible improvements.

2.4.3. *Weighted CLOE method*

One of the basic CLOE method caveats is the fact that, during the successive updating, the history of the process dynamics tends to overinfluence the current results. As such, a first improvement of this method involves weighting the contribution of the old I/O data, compared with the contribution of the recently acquired data. The method presented in this section is referred to as *weighted CLOE* (W-CLOE).

Unlike the basic CLOE method, the starting equation used to build the updating process in the W-CLOE method is:

$$\hat{\boldsymbol{\theta}}_N = \left(\alpha_{N-1} \sum_{n=1}^{N-1} \boldsymbol{\varphi}[n]\boldsymbol{\varphi}^T[n] + \beta_{N-1}\boldsymbol{\varphi}[N]\boldsymbol{\varphi}^T[N] \right)^{-1} \cdot \left(\alpha_{N-1} \sum_{n=1}^{N-1} \boldsymbol{\varphi}[n]y[n] + \beta_{N-1}\boldsymbol{\varphi}[N]y[N] \right) \quad [2.66]$$

where $\alpha_{N-1} \in (0,1]$ and $\beta_{N-1} \in [1,2]$ are adaptive weights (as defined later). The role of weight α_{N-1} is to attenuate the effect of the process dynamics history, while the weight β_{N-1} is employed to highlight the importance of the current I/O data.

Following the reasoning of online LSM, the main matrix is defined as follows:

$$\mathbf{P}_N = \left(\alpha_{N-1} \sum_{n=1}^{N-1} \boldsymbol{\varphi}[n]\boldsymbol{\varphi}^T[n] + \beta_{N-1}\boldsymbol{\varphi}[N]\boldsymbol{\varphi}^T[N] \right)^{-1} \cong \left(\alpha_{N-1}\mathbf{P}_{N-1}^{-1} + \beta_{N-1}\boldsymbol{\varphi}[N]\boldsymbol{\varphi}^T[N] \right)^{-1} \quad [2.67]$$

Note that the recursive equation [2.67] is not exact, because the matrix $\mathbf{P}_{N-1}^{-1}$ equals the sum $\sum_{i=1}^{N-1} \boldsymbol{\varphi}[i]\boldsymbol{\varphi}^T[i]$ for $\alpha_{N-2} = \beta_{N-2} = 1$ only. The farther α_{N-2} or β_{N-2} is from the unit value, the less accurate the expression [2.67]. In the sequel, one considers, however, that equation [2.67] acts like an identity.

By applying Sherman–Morrison's Lemma to equation [2.67], one obtains:

$$\mathbf{P}_N = \frac{1}{\alpha_{N-1}}\left(\mathbf{P}_{N-1} - \frac{\beta_{N-1}\mathbf{P}_{N-1}\boldsymbol{\varphi}[N]\boldsymbol{\varphi}^T[N]\mathbf{P}_{N-1}}{\alpha_{N-1} + \beta_{N-1}\boldsymbol{\varphi}^T[N]\mathbf{P}_{N-1}\boldsymbol{\varphi}[N]}\right). \quad [2.68]$$

Also, equation [2.67] involves another interesting recurrent relation:

$$\mathbf{P}_{N-1}^{-1} = \frac{1}{\alpha_{N-1}}\left(\mathbf{P}_N^{-1} - \beta_{N-1}\boldsymbol{\varphi}[N]\boldsymbol{\varphi}^T[N]\right). \quad [2.69]$$

At the current instant $n \geq 1$, one can note that:

$$\hat{\boldsymbol{\theta}}_n = \mathbf{P}_n\left(\alpha_{n-1}\sum_{i=1}^{n-1}\boldsymbol{\varphi}[i]y[i] + \beta_{n-1}\boldsymbol{\varphi}[n]y[n]\right) \cong \mathbf{P}_n\left(\alpha_{n-1}\mathbf{P}_{n-1}^{-1}\hat{\boldsymbol{\theta}}_{n-1} + \beta_{n-1}\boldsymbol{\varphi}[n]y[n]\right)$$

$$= \mathbf{P}_n\left[\cancel{\alpha_{n-1}}\frac{1}{\cancel{\alpha_{n-1}}}\left(\mathbf{P}_n^{-1} - \beta_{n-1}\boldsymbol{\varphi}[n]\boldsymbol{\varphi}^T[n]\right)\hat{\boldsymbol{\theta}}_{n-1} + \beta_{n-1}\boldsymbol{\varphi}[n]y[n]\right]$$

$$= \hat{\boldsymbol{\theta}}_{n-1} + \beta_{n-1}\mathbf{P}_n\boldsymbol{\varphi}[n]\underbrace{\left(y[n] - \boldsymbol{\varphi}^T[n]\hat{\boldsymbol{\theta}}_{n-1}\right)}_{\varepsilon_y^b[n]} = \hat{\boldsymbol{\theta}}_{n-1} + \beta_{n-1}\mathbf{P}_n\boldsymbol{\varphi}[n]\varepsilon_y^b[n], \ \forall n \geq 1. \quad [2.70]$$

As previously, equation [2.70] is not exact because the term $\mathbf{P}_{n-1}^{-1}\hat{\boldsymbol{\theta}}_{n-1}$ equals the sum $\sum_{i=1}^{n-1}\boldsymbol{\varphi}[i]y[i]$ for $\alpha_{n-2} = \beta_{n-2} = 1$ only. The farther α_{n-2} or β_{n-2} from the unit value, the less accurate equation [2.70]. In the sequel, one considers, however, that equation [2.70] acts like an identity.

The recipe [2.70] to update the parameters of the identification model is only slightly different from the one used in the context of the online LSM. One can continue the reasoning in order to remove the presence of the current main matrix ($\mathbf{P}_n$) in the correcting term (which makes implementation difficult). The result of inversion lemma [2.68] is employed to write that:

$$\mathbf{P}_n\boldsymbol{\varphi}[n] = \frac{1}{\alpha_{n-1}}\left(\mathbf{P}_{n-1}\boldsymbol{\varphi}[n] - \frac{\beta_{n-1}\mathbf{P}_{n-1}\boldsymbol{\varphi}[n]\boldsymbol{\varphi}^T[n]\mathbf{P}_{n-1}\boldsymbol{\varphi}[n]}{\alpha_{n-1} + \beta_{n-1}\boldsymbol{\varphi}^T[n]\mathbf{P}_{n-1}\boldsymbol{\varphi}[n]}\right)$$

$$= \frac{1}{\cancel{\alpha_{n-1}}}\,\frac{\cancel{\alpha_{n-1}}\mathbf{P}_{n-1}\boldsymbol{\varphi}[n] + \cancel{\beta_{n-1}\mathbf{P}_{n-1}\boldsymbol{\varphi}[n]\boldsymbol{\varphi}^T[n]\mathbf{P}_{n-1}\boldsymbol{\varphi}[n]} - \cancel{\beta_{n-1}\mathbf{P}_{n-1}\boldsymbol{\varphi}[n]\boldsymbol{\varphi}^T[n]\mathbf{P}_{n-1}\boldsymbol{\varphi}[n]}}{\alpha_{n-1} + \beta_{n-1}\boldsymbol{\varphi}^T[n]\mathbf{P}_{n-1}\boldsymbol{\varphi}[n]} \quad [2.71]$$

$$= \frac{\mathbf{P}_{n-1}\boldsymbol{\varphi}[n]}{\alpha_{n-1} + \beta_{n-1}\boldsymbol{\varphi}^T[n]\mathbf{P}_{n-1}\boldsymbol{\varphi}[n]}.$$

From [2.70] and [2.71], it results:

$$\hat{\boldsymbol{\theta}}_n = \hat{\boldsymbol{\theta}}_{n-1} + \beta_{n-1}\mathbf{P}_n\boldsymbol{\varphi}[n]\varepsilon_y^b[n] = \hat{\boldsymbol{\theta}}_{n-1} + \frac{\beta_{n-1}\mathbf{P}_{n-1}\boldsymbol{\varphi}[n]}{\alpha_{n-1} + \beta_{n-1}\boldsymbol{\varphi}^T[n]\mathbf{P}_{n-1}\boldsymbol{\varphi}[n]}\varepsilon_y^b[n], \quad [2.72]$$

which allows for an efficient implementation, since the matrix $\mathbf{P}_n$ (at the current instant) is replaced by an expression that contains the previous matrix, $\mathbf{P}_{n-1}$.

Moreover, as in the case of the basic CLOE method, there is a correlation between the two output errors:

$$\begin{aligned}\varepsilon_y^f[n] &= y[n] - \boldsymbol{\varphi}^T[n]\hat{\boldsymbol{\theta}}_n = \underbrace{y[n] - \boldsymbol{\varphi}^T[n]\hat{\boldsymbol{\theta}}_{n-1}}_{\varepsilon_y^b[n]} - \frac{\beta_{n-1}\boldsymbol{\varphi}^T[n]\mathbf{P}_{n-1}\boldsymbol{\varphi}[n]}{\alpha_{n-1} + \beta_{n-1}\boldsymbol{\varphi}^T[n]\mathbf{P}_{n-1}\boldsymbol{\varphi}[n]}\varepsilon_y^b[n] \\ &= \frac{\alpha_{n-1} + \cancel{\beta_{n-1}\boldsymbol{\varphi}^T[n]\mathbf{P}_{n-1}\boldsymbol{\varphi}[n]} - \cancel{\beta_{n-1}\boldsymbol{\varphi}^T[n]\mathbf{P}_{n-1}\boldsymbol{\varphi}[n]}}{\alpha_{n-1} + \beta_{n-1}\boldsymbol{\varphi}^T[n]\mathbf{P}_{n-1}\boldsymbol{\varphi}[n]}\varepsilon_y^b[n] \\ &= \frac{\alpha_{n-1}}{\alpha_{n-1} + \beta_{n-1}\boldsymbol{\varphi}^T[n]\mathbf{P}_{n-1}\boldsymbol{\varphi}[n]}\varepsilon_y^b[n].\end{aligned} \quad [2.73]$$

(In order to make the writing simpler, for the moment, no estimated regressors vector has been employed.)

To complete the reasoning, the implementation formula for the main updating relation [2.72] can be written as follows (with property [2.73]):

$$\hat{\boldsymbol{\theta}}_n = \hat{\boldsymbol{\theta}}_{n-1} + \frac{\beta_{n-1}\mathbf{P}_{n-1}\boldsymbol{\varphi}[n]}{\alpha_{n-1} + \beta_{n-1}\boldsymbol{\varphi}^T[n]\mathbf{P}_{n-1}\boldsymbol{\varphi}[n]}\varepsilon_y^b[n] = \hat{\boldsymbol{\theta}}_{n-1} + \frac{\beta_{n-1}}{\alpha_{n-1}}\mathbf{P}_{n-1}\boldsymbol{\varphi}[n]\varepsilon_y^f[n]. \quad [2.74]$$

How can the weightings α_n and β_n be updated? A first option is to set them to constant values, empirically, following an analysis of the process dynamics. For example, α_n can be set to 0.9 and β_n to 1.1. This choice lacks finesse, however. Moreover, the optimal digital adaptation procedure must not degrade the consistency of the estimated parameters. There is a theoretical result that introduces sufficient conditions to preserve the consistency of estimations $\left\{\hat{\boldsymbol{\theta}}_n\right\}_{n\geq 0}$.

THEOREM 2.3. *In the context of W-CLOE method, the series of estimations* $\left\{\hat{\boldsymbol{\theta}}_n\right\}_{n\geq 0}$ *updated with the recursive equation [2.74] is consistent if the transfer function*:

$$\mathrm{H_{S,P}}\left(z^{-1}\right) = \frac{\mathrm{S}\left(z^{-1}\right)}{\mathrm{P}\left(z^{-1}\right)} - \frac{\beta_{\max}}{2},$$

with

$$\beta_{\max} = \sup_{n\geq 0}\{\beta_n\} \in (0,2),$$

is strictly real positive.

Recall that the transfer function of a continuous system, say H_c, is strictly real positive if:

$$\begin{cases} \mathrm{Re}[H_c(s)] > 0, & \text{if } \mathrm{Re}(s) > 0 \\ \mathrm{Im}[H_c(s)] = 0, & \text{if } \mathrm{Im}(s) = 0 \end{cases}. \qquad [2.75]$$

In the case of a discrete system, the transfer function, say H_d, is strictly real positive if:

$$\begin{cases} \mathrm{Re}\left[H_d\left(z^{-1}\right)\right] > 0, & \text{if } |z| > 1 \\ \mathrm{Im}\left[H_d\left(z^{-1}\right)\right] = 0, & \text{if } \arg(z) = 0 \end{cases}. \qquad [2.76]$$

In Theorem 2.3, the global polynomial P provides the poles of the CL system. The expression of the transfer function $H_{S,P}$ justifies the choice of upper limit for weights β_n (equal to 2). If all weights α_n and β_n are unit, then the basic CLOE procedure is obtained. In this case, $\beta_{\max} = 1$ and it can be shown that, if the RST regulator is designed properly, then the transfer function:

$$H_{S,P}\left(z^{-1}\right) = \frac{S\left(z^{-1}\right)}{P\left(z^{-1}\right)} - \frac{1}{2} \qquad [2.77]$$

is strictly real positive. This means that the basic CLOE method produces consistent estimations of the parameters vector in the identification model. The problem is, therefore, to find a strategy for increasing the weight β_n up to the positiveness limit of the transfer function $H_{S,P}$.

In fact, Theorem 2.3 offers the possibility of testing at any instant whether the choice of the weight β_n is correct or not. To do this, the maximal current value of the series $\{\beta_i\}_{i\in\overline{0,n}}$ is calculated, and the positiveness of the current transfer function $H_{S,P}\left(z^{-1}\right)$ is tested. Note that this function can change from one sampling step to another, as the RST regulator has to be updated as well.

Nevertheless, the problem of selecting the weights α_n and β_n still remains, and a numerical procedure is needed to solve it. Another approach (different from the empirical one) is to analyze the frequency of the estimated behavior. More precisely, the idea is to focus on the bandwidths around the critical points (provided by the intrinsic resonances of the identification model). In this case, the weights have to be varied in order to reduce (if possible) the resonant spectral lines, as illustrated in Figure 2.4. In this figure, the spectrum of the identification model (drawn in *decibels* [dB]) is evaluated using the *Fourier transform* [OPP 85a, OPP 85b, PRO 96], which has been applied to the estimated transfer function:

$$\hat{H}_n\left(e^{j\omega}\right) = e^{-j\omega k}\frac{\hat{B}_n\left(e^{j\omega}\right)}{\hat{A}_n\left(e^{j\omega}\right)}, \quad \forall \omega \in [0,\pi]. \qquad [2.78]$$

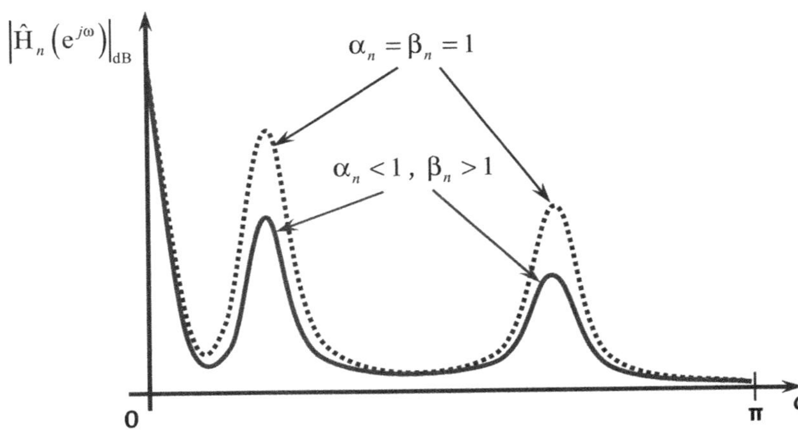

Figure 2.4. *Attenuation of resonant spectral lines for a process by varying the weights in W-CLOE method*

There are cases where the model of the process decreases its resonances following an astute choice of adaptive weights. Since this attenuation can be quantified in such a way that an optimization criterion is defined, one rather seeks optimal weights. Nevertheless, optimizing this criterion using classical techniques is not easy, as its derivatives are difficult (or even impossible) to calculate. In turn, a meta-heuristic solution to the optimization problem can be found. In [STE 14], some meta-heuristic optimization methods are suggested (based on the Monte Carlo principle). The reader may design a numerical procedure derived from this group of methods in order to solve the optimization problem (since such procedures are quite simple and do not raise the complexity of W-CLOE algorithm).

Another interesting approach is linked to the renewal equation [2.74]. By carefully looking at this recursive relation, one notices its similarity to the core equation in a numerical optimization procedure based on exact methods. For example, in [BOR 13c] or [BOR 13d], there are two nonlinear programming methods based on recursive relations like [2.74], namely the *Cauchy method* (also known as gradient-based method) and the *Newton–Raphson method*, both using variable adaptation steps. Recall the main updating equations of each method.

a) For the Cauchy method:

$$\begin{cases} \mathbf{x}_n = \mathbf{x}_{n-1} + \alpha_{n-1}\mathbf{J}_{\mathbf{x}}(\mathbf{x}_{n-1}) \\ \alpha_n = \alpha_{n-1} - \mathbf{J}_{\mathbf{x}}^T(\mathbf{x}_n)\mathbf{J}_{\mathbf{x}}(\mathbf{x}_{n-1}) \end{cases}, \quad \forall n \geq 1. \qquad [2.79]$$

b) For the Newton–Raphson method:

$$\begin{cases} \mathbf{x}_n = \mathbf{x}_{n-1} + \alpha_{n-1}\mathbf{J}_{\mathbf{xx}}^{-1}(\mathbf{x}_{n-1})\mathbf{J}_{\mathbf{x}}(\mathbf{x}_{n-1}) \\ \alpha_n = \alpha_{n-1} - \dfrac{\mathbf{J}_{\mathbf{x}}^T(\mathbf{x}_n)\mathbf{J}_{\mathbf{xx}}^{-1}(\mathbf{x}_{n-1})\mathbf{J}_{\mathbf{x}}(\mathbf{x}_{n-1})}{\mathbf{J}_{\mathbf{x}}^T(\mathbf{x}_{n-1})\mathbf{J}_{\mathbf{xx}}^{-1}(\mathbf{x}_{n-1})\mathbf{J}_{\mathbf{xx}}(\mathbf{x}_n)\mathbf{J}_{\mathbf{xx}}^{-1}(\mathbf{x}_{n-1})\mathbf{J}_{\mathbf{x}}(\mathbf{x}_{n-1})} \end{cases}, \quad \forall n \geq 1. \qquad [2.80]$$

In systems [2.79] and [2.80]:

– J is the optimization criterion, with the gradient vector denoted by $\mathbf{J}_{\mathbf{x}}$ and the Hessian matrix denoted by $\mathbf{J}_{\mathbf{xx}}$;

– $\mathbf{x}_{n-1}$ is the current optimal point;

– α_{n-1} is the current adaptation step.

Therefore, in equation [2.74], one can consider the ratio β_{n-1}/α_{n-1} to be the adaptation step, $\boldsymbol{\theta}_n$ to be $\mathbf{x}_n$ and:

a) either $\mathbf{P}_{n-1}\boldsymbol{\varphi}[n]\varepsilon_y^f[n]$ corresponds to the current gradient ($\mathbf{J}_{\mathbf{x}}(\mathbf{x}_{n-1})$) of a given criterion;

b) or $\boldsymbol{\varphi}[n]\varepsilon_y^f[n]$ corresponds to the current gradient ($\mathbf{J}_{\mathbf{x}}(\mathbf{x}_{n-1})$) and $\mathbf{P}_{n-1}^{-1}$ (symmetrical, invertible) corresponds to the current Hessian matrix ($\mathbf{J}_{\mathbf{xx}}(\mathbf{x}_{n-1})$) of another criterion.

In any case, even if the optimization criterion is unknown, the successive values of its gradient and, potentially, of its Hessian matrix are available at each sampling instant. It is therefore normal to use the second equation of systems [2.79] and [2.80], in order to update the ratio β_n / α_n. More specifically:

a) for the Cauchy method:

$$\frac{\beta_n}{\alpha_n} = \frac{\beta_{n-1}}{\alpha_{n-1}} - \boldsymbol{\varphi}^T[n+1]\mathbf{P}_n\mathbf{P}_{n-1}\boldsymbol{\varphi}[n]\varepsilon_y^f[n+1]\varepsilon_y^f[n], \quad \forall n \geq 1; \qquad [2.81]$$

b) for the Newton–Raphson method:

$$\frac{\beta_n}{\alpha_n} = \frac{\beta_{n-1}}{\alpha_{n-1}} - \frac{\boldsymbol{\varphi}^T[n+1]\mathbf{P}_{n-1}\boldsymbol{\varphi}[n]\varepsilon_y^f[n+1]\varepsilon_y^f[n]}{\boldsymbol{\varphi}^T[n]\mathbf{P}_{n-1}\mathbf{P}_n^{-1}\mathbf{P}_{n-1}\boldsymbol{\varphi}[n]\left(\varepsilon_y^f[n]\right)^2}, \quad \forall n \geq 1. \qquad [2.82]$$

Note that the factor $\mathbf{P}_n^{-1}$ (from the denominator in equation [2.82]) does not require the inversion of a matrix at every sampling instant (once again, see equation [2.67]). All that is needed is to invert the matrix $\mathbf{P}_0$ in the beginning of the numerical procedure and to then update the obtained inverse through recursive relation [2.67].

The main problem raised by equations [2.81] and [2.82] concerns the effective evaluation of *a posteriori* error at the next instant ($\varepsilon_y^f[n+1]$). Obviously, due to definition [2.62], the next value of this error cannot be evaluated before acquiring the next output value ($y[n+1]$), which makes the implementation impossible. All that can be done is to approximate this error. A natural approximation (especially in the vicinity of optimum) is $\varepsilon_y^f[n+1] \cong \varepsilon_y^f[n]$, which would allow implementing equations [2.81] and [2.82]. Moreover, the recursive equation [2.82] becomes independent of the *a posteriori* error, as the factor $\left(\varepsilon_y^f[n]\right)^2$ is totally cancelled.

In order to separately specify the weights, either the following constraint is imposed:

$$\alpha_n + \beta_n = 2, \quad \forall n \geq 0, \qquad [2.83]$$

or, as in the case of equation [2.82], one can go as far as to consider that:

$$\begin{cases} \alpha_n = \alpha_{n-1} + \boldsymbol{\varphi}^T[n]\mathbf{P}_{n-1}\mathbf{P}_n^{-1}\mathbf{P}_{n-1}\boldsymbol{\varphi}[n] \\ \beta_n = \beta_{n-1} - \boldsymbol{\varphi}^T[n+1]\mathbf{P}_{n-1}\boldsymbol{\varphi}[n] \end{cases}, \quad \forall n \geq 1. \qquad [2.84]$$

This approach does contain an important caveat though. The weights vary within given limits: $\alpha_n \in (0,1]$ and $\beta_n \in [1,2)$ for any $n \geq 0$. If the natural constraint [2.83] is imposed, then the weights can be expressed as: $\alpha_n = 1-\delta_n$ and $\beta_n = 1+\delta_n$, where the variable offset δ_n moves between 0 and 1. Using the weights' new expressions, it is possible to go back to the recurrences [2.81] and [2.82] and write that:

$$\frac{1+\delta_n}{1-\delta_n} = \frac{1+\delta_{n-1}}{1-\delta_{n-1}} + \Delta_{n-1}, \quad \forall n \geq 1, \qquad [2.85]$$

where the notation Δ_n is employed to unify the two possible corrections:

$$\Delta_n \in \left\{ -\boldsymbol{\varphi}^T[n+1]\mathbf{P}_n\mathbf{P}_{n-1}\boldsymbol{\varphi}[n]\left(\varepsilon_y^f[n]\right)^2, -\frac{\boldsymbol{\varphi}^T[n+1]\mathbf{P}_{n-1}\boldsymbol{\varphi}[n]}{\boldsymbol{\varphi}^T[n]\mathbf{P}_{n-1}\mathbf{P}_n^{-1}\mathbf{P}_{n-1}\boldsymbol{\varphi}[n]} \right\}, \quad \forall n \geq 1. \qquad [2.86]$$

From [2.85], after a few elementary algebraic manipulations, one obtains the recurrence verified by the offset:

$$\delta_n = \frac{2\delta_{n-1} + (1-\delta_{n-1})\Delta_{n-1}}{2 + (1-\delta_{n-1})\Delta_{n-1}}, \quad \forall n \geq 1. \qquad [2.87]$$

It can be shown (by mathematical induction) that, if the initial offset varies between 0 and 1, and no correction Δ_n is located in the interval $\left[-2/(1-\delta_n), -2\delta_n/(1-\delta_n)\right]$, then all of the successive offsets also vary between 0 and 1. Since it cannot be shown that Δ_n verifies this constraint, one has to adjust it as soon as its current value gets close to the forbidden interval.

To conclude this section, the corresponding numerical procedure of W-CLOE method is described in Algorithm 2.3.

As one can see, in Algorithm 2.3, the weights updating technique described in equations [2.87] and [2.82] has been adopted. However, the user can propose other W-CLOE algorithms, as the procedure above is founded on the inexact recurrent equations [2.67] and [2.70].

Obviously, Algorithm 2.2 (basic CLOE) can easily be obtained from the Algorithm 2.3 (W-CLOE), simply by keeping all of the offsets δ_n at their initial null value. Consequently, the methods described in the remainder of this chapter are derived from the basic CLOE method, with the additional possibility of adding some weights.

1) *Initialization*:

a) Set the structural index of the model: $n\theta = na + nb \in \mathbb{N}^*$.

b) Identify a CL model of the process using one of the offline techniques suggested in the previous section. As a result, a first estimation of five polynomials is available: $\hat{A}_0$, $\hat{B}_0$, $\hat{R}_0$, $\hat{S}_0$ and $\hat{T}_0$. Estimations of the process pure delay (k) and of the command ($\hat{u}$) are also available. This allows building the initial vector of estimated parameters, $\hat{\boldsymbol{\theta}}_0$, by using the coefficients of polynomials $\hat{A}_0$ and $\hat{B}_0$.

c) Choose an initialization for the main matrices: $\mathbf{P}_{-1} = \mathbf{P}_0 = \alpha^2 \mathbf{I}_{n\theta}$, with a large enough value of α (a few hundreds, at least).

d) Calculate the inverse of the main matrix (if necessary): $\mathbf{P}_0^{-1} = \dfrac{1}{\alpha^2}\mathbf{I}_{n\theta}$.

e) Initialize the vector of estimated regressors ($\hat{\boldsymbol{\varphi}}[0]$ and $\hat{\boldsymbol{\varphi}}[1]$) using the most recent I/O data, acquired for the offline identification (see step 1(d) of the previous algorithm, for example).

f) Initialize the weights offset: $\delta_0 = 0$ (leading to $\alpha_0 = \beta_0 = 1$).

g) Set a correction threshold: $\varepsilon > 0$ (by default, $\varepsilon = 0.1$).

h) Set the starting iteration index: $n = 1$.

2) *For* $n \geq 1$:

2.1. Acquire the current reference and output: $\{y_r[n], y[n]\}$.

2.2. Estimate the current control: $\hat{u}[n] = \dfrac{\hat{T}_{n-1}(q^{-1})}{\hat{S}_{n-1}(q^{-1})} y_r[n] - \dfrac{\hat{R}_{n-1}(q^{-1})}{\hat{S}_{n-1}(q^{-1})} \hat{\boldsymbol{\varphi}}^T[n]\hat{\boldsymbol{\theta}}_{n-1}$.

2.3. Prepare the regressors vector for the next iteration

$$\hat{\boldsymbol{\varphi}}[n+1] = [-y[n] \quad \cdots \quad -y[n-na+1] \mid \hat{u}[n-k+1] \quad \cdots \quad \hat{u}[n-nb-k+1]]^T.$$

(Take values from vector $\hat{\boldsymbol{\varphi}}[1]$, if necessary.)

2.4. Evaluate the *a priori* output error: $\varepsilon_y^b[n] = y[n] - \hat{\boldsymbol{\varphi}}^T[n]\hat{\boldsymbol{\theta}}_{n-1}$.

2.5. Evaluate the auxiliary vector: $\hat{\boldsymbol{\xi}}_n = \mathbf{P}_{n-1}\hat{\boldsymbol{\varphi}}[n]$.

2.6. Evaluate the multiplicity constant of the output errors:

$$\hat{\mu}_{n-1} = \frac{1}{1-\delta_{n-1} + (1+\delta_{n-1})\hat{\boldsymbol{\varphi}}^T[n]\hat{\boldsymbol{\xi}}_n} \quad \text{(scalar).}$$

2.7. Evaluate the *a posteriori* output error: $\varepsilon_y^f[n] = (1-\delta_{n-1})\hat{\mu}_{n-1}\varepsilon_y^b[n]$.

2.8. Update the main matrix: $\mathbf{P}_n = \dfrac{1}{1-\delta_{n-1}}\left[\mathbf{P}_{n-1} - (1+\delta_{n-1})\hat{\mu}_{n-1}\hat{\boldsymbol{\xi}}_n\hat{\boldsymbol{\xi}}_n^T\right]$.

2.9. Update the main matrix inverse (if necessary):

$$\mathbf{P}_n^{-1} = (1-\delta_{n-1})\mathbf{P}_{n-1}^{-1} + (1+\delta_{n-1})\hat{\boldsymbol{\varphi}}[n]\hat{\boldsymbol{\varphi}}^T[n].$$

2.10. Update the parameters vector: $\hat{\boldsymbol{\theta}}_n = \hat{\boldsymbol{\theta}}_{n-1} + \frac{1+\delta_{n-1}}{1-\delta_{n-1}}\hat{\xi}_n \varepsilon_y^f[n]$.

2.11. Update the RST regulator: $\{\hat{R}_n, \hat{S}_n, \hat{T}_n\}$ from the model $\{\hat{A}_n, \hat{B}_n\}$ (built from $\hat{\boldsymbol{\theta}}_n$).

2.12. Calculate the correction to be applied to the weights ratio:

$$\Delta_{n-1} = \hat{\boldsymbol{\varphi}}^T[n+1]\mathbf{P}_n\mathbf{P}_{n-1}\hat{\boldsymbol{\varphi}}[n]\left(\varepsilon_y^f[n]\right)^2 \text{ or } \Delta_{n-1} = -\frac{\hat{\boldsymbol{\varphi}}^T[n+1]\mathbf{P}_{n-1}\hat{\boldsymbol{\varphi}}[n]}{\hat{\boldsymbol{\varphi}}^T[n]\mathbf{P}_{n-1}\mathbf{P}_n^{-1}\mathbf{P}_{n-1}\hat{\boldsymbol{\varphi}}[n]}.$$

2.13. If $\Delta_{n-1} \in \left[\frac{-2}{1-\delta_{n-1}}, \frac{-1-\delta_{n-1}}{1-\delta_{n-1}}\right)$, enforce the correction to leave the interval on the left side: $\Delta_{n-1} \leftarrow \frac{-2}{1-\delta_{n-1}} - \varepsilon$.

2.14. If $\Delta_{n-1} \in \left[\frac{-1-\delta_{n-1}}{1-\delta_{n-1}}, \frac{-2\delta_{n-1}}{1-\delta_{n-1}}\right]$, enforce the correction to leave the interval on the right side: $\Delta_{n-1} \leftarrow \frac{-2\delta_{n-1}}{1-\delta_{n-1}} + \varepsilon$.

2.15. Update the weights offset (for the next iteration):

$$\delta_n = \frac{2\delta_{n-1} + (1-\delta_{n-1})\Delta_{n-1}}{2 + (1-\delta_{n-1})\Delta_{n-1}}.$$

2.16. Evaluate/update the maximum value of $\{\delta_i\}_{i\in\overline{0,n}}$, namely $\delta_n^{\max}$.

2.17. If the transfer function:

$$H_{\hat{S}_n,\hat{P}_n}(z^{-1}) = \frac{\hat{S}_n(z^{-1})}{\hat{P}_n(z^{-1})} - \frac{1+\delta_n^{\max}}{2},$$

where $\hat{P}_n(z^{-1}) = \hat{A}_n(z^{-1})\hat{S}_n(z^{-1}) + z^{-k}\hat{B}_n(z^{-1})\hat{R}_n(z^{-1})$, is not real positive, progressively decrease the offset δ_n (for example, by using multiple of ε as decrements) until the property is verified.

2.18. Move on to the next iteration: $n \leftarrow n+1$.

3) *Return the model of CL process* $\{\hat{A}_n, \hat{B}_n, \hat{R}_n, \hat{S}_n, \hat{T}_n\}$ *at every instant* $n \geq 0$. *An estimation of the command* $\hat{u}$ *is also available.*

Algorithm 2.3. *Steps of W-CLOE procedure*

2.4.4. *Filtered CLOE method or adaptively filtered CLOE*

For some processes, the sufficient condition of consistence proposed in Theorem 2.3 can prove itself difficult to verify. In this case, it is better to use a different version of the CLOE method, making the task easier. This is how the *CLOE method with filtered regressors vector*, in short F-CLOE (*filtered CLOE*), appeared.

The role of the filter applied to the measured I/O data is to attenuate the disturbances and therefore to increase the *signal-to-noise ratio (SNR)*, especially when these disturbances excessively degrade the useful data expressing the process dynamics. There are several possibilities for choosing this filter (or for designing it). Normally, a specific filter corresponds to each process. In the scientific literature, there is a relatively rich selection of design methods for analog or digital filters (for example, see [OPP 85a] or [PRO 96]). Sometimes, it is not easy to find the "right filter", as the user is faced with the difficulty of defining the "noise" in a numerical way.

Usually, in SI, the data filters used are strongly linked to the identification models that can be obtained. As such, the F-CLOE method proposes the use of the following filter:

$$F\left(q^{-1}\right)=\frac{\hat{S}\left(q^{-1}\right)}{\hat{P}_0\left(q^{-1}\right)}, \qquad [2.88]$$

where $\hat{P}_0$ is an offline estimation of polynomial [2.7], which gives CL system poles. Also, the polynomial $\hat{S}$ is an estimation of polynomial that defines the IIR filter [2.2] integrated into the RST regulator.

To determine the polynomial $\hat{P}_0$, the CL process has to be seen as a black box that can be identified using the OLSI techniques. Normally, the class of employed models is ARMAX. The ARX model (that belongs to this class) is often targeted as part of this approach. Finally, due to the increase in power of modern computers, *Box-Jenkins* (BJ) type models also are an interesting choice (or even recommended). (A description of linear identification models can be found, for instance, in [BOR 13a] or [BOR 13b].)

One has to outline that the input to this global black box is the reference signal y_r and not the control signal u. If the CL process supports PRS as references (adapted to the variation range of output y), then a relatively accurate estimation of the polynomial P can be obtained, without knowing the other polynomials of definition [2.7]. Concerning the estimation $\hat{S}$, several approaches are available. For example, one can consider that $\hat{S}$ is identical to $\hat{S}_0$, as obtained during the initialization step of CLOE class algorithms (previously described). In this case, the filter [2.88] is constant and not adaptive. To make it adaptive, $\hat{S}$ can be replaced by $\hat{S}_n$, obtained during the RST regulator update step within the CLOE procedures. In any case, the resulting filter F is invertible, as the polynomials S are all stable (since they are returned by the procedure of RST regulator design/updating). If the filter is adaptive, the method is referred to as *adaptively filtered CLOE* (AF-CLOE).

The numerical algorithm corresponding to the (A)F-CLOE methods is very similar to Algorithms 2.2 and 2.3. There are, however, three differences:

a) The initialization step has to be completed using the offline identification of the global black box. The identification model is ARX, ARMAX or BJ (the latter being recommended). Nevertheless, independently of the chosen model, the useful filter is expressed using a transfer function whose denominator is the polynomial $\hat{P}_0$. This operation can be carried out using OLSI techniques (such as those described in [STE 05, BOR 13a] or [BOR 13b]).

b) After having prepared the regressors vector for the next iteration (see step 2.3 of previous CLOE algorithms), a filtered version of this vector has to be evaluated and stored in memory:

$$\hat{\varphi}_F[n+1] = F\left(q^{-1}\right)\hat{\varphi}[n+1] = \frac{\hat{S}_n\left(q^{-1}\right)}{\hat{P}_0\left(q^{-1}\right)}\hat{\varphi}[n+1]\,. \qquad [2.89]$$

Note that the command is estimated using the (unfiltered) vector $\hat{\varphi}[n]$, while, for all other steps of the algorithm, $\hat{\varphi}$ is replaced by $\hat{\varphi}_F$. As a result, the filtered vector must not overwrite the unfiltered one in the computer program, as both vectors are required to reach the goal.

c) The condition of consistency changes in the same way. The transfer function $H_{\hat{S}_n,\hat{P}_n}$ (from step 2.17 of Algorithm 2.3) has to be replaced by:

$$H_{\hat{P}_0,\hat{P}_n} = \frac{\cancel{\hat{S}_n\left(z^{-1}\right)}}{\hat{P}_n\left(z^{-1}\right)}\frac{\hat{P}_0\left(q^{-1}\right)}{\cancel{\hat{S}_n\left(q^{-1}\right)}} - \frac{1+\delta_{n+1}^{\max}}{2} = \frac{\hat{P}_0\left(q^{-1}\right)}{\hat{P}_n\left(q^{-1}\right)} - \frac{1+\delta_{n+1}^{\max}}{2} \qquad [2.90]$$

(this expression being obtained by applying the inverse of the data filter).

The filter definition [2.88] relies on the following result of consistency.

THEOREM 2.4. *In the context of (A)F-CLOE methods, the updated parameters vector estimations* $\left\{\hat{\boldsymbol{\theta}}_n\right\}_{n\geq 0}$ *is consistent if the transfer function*:

$$\mathrm{H}_{\mathrm{S,P}}\left(z^{-1}\right) = \frac{\hat{\mathrm{P}}_0\left(z^{-1}\right)}{\mathrm{P}\left(z^{-1}\right)} - \frac{\beta_{\max}}{2},$$

with

$$\beta_{\max} = \sup_{n\geq 0}\{\beta_n\} \in (0,2),$$

is strictly real positive.

Another approach that provides numerical procedures of AF-CLOE class is based on the periodic reidentification in OL of polynomial $\hat{\mathrm{P}}_0$ (for instance, with a delay of 10 sampling periods). The I/O data previously acquired during the online procedure are employed in this aim.

In some publications, the (A)F-CLOE methods (and especially AF-CLOE) have resulted in greater levels of performance than the basic CLOE, notably in critical frequency areas.

2.4.5. *Extended CLOE method*

The use of ARX models is starting to lose popularity in modern applications of SI. The model emerged around 30 years ago, mainly because of the fact that the corresponding regressors vector only contains acquired I/O data (and not signals that need to be estimated). This makes the digital identification procedure (usually LSM) very efficient, with a considerable reduction in the computational burden. However, this model contains two main limitations: it imposes a disturbance filter that contains no zeros and, moreover, this filter has exactly the same poles as the useful filter. These limitations clearly appear in the main equation of the model:

$$y[n] = \underbrace{\mathrm{q}^{-k}\frac{\mathrm{B}\left(\mathrm{q}^{-1}\right)}{\mathrm{A}\left(\mathrm{q}^{-1}\right)}}_{\text{useful filter}}u[n] + \underbrace{\frac{1}{\mathrm{A}\left(\mathrm{q}^{-1}\right)}}_{\text{disturbances filter}}e[n], \quad \forall n\in\mathbb{N}, \qquad [2.91]$$

where e is white noise (usually centered and Gaussian, following the Central Limit Theorem). Looking at Figure 2.3, one can see that, in the lower part, there is an estimation of the process disturbance, $\hat{v}$. The ARX model proposes estimating it by filtering the white noise:

$$\hat{v}[n] = \frac{1}{\hat{\mathrm{A}}\left(\mathrm{q}^{-1}\right)} e[n], \quad \forall n \in \mathbb{N}, \qquad [2.92]$$

which clearly shows the above mentioned limitations.

In order to improve the accuracy of identification procedure in CL, other more complex models of disturbance can be considered, even though this results in heftier computations. Two models are considered in this aim:

$$-\text{ ARMAX: } v[n] = \frac{\mathrm{C}\left(\mathrm{q}^{-1}\right)}{\mathrm{A}\left(\mathrm{q}^{-1}\right)} e[n], \quad \forall n \in \mathbb{N}; \qquad [2.93]$$

$$-\text{ BJ: } v[n] = \frac{\mathrm{C}\left(\mathrm{q}^{-1}\right)}{\mathrm{D}\left(\mathrm{q}^{-1}\right)} e[n], \quad \forall n \in \mathbb{N}. \qquad [2.94]$$

The ARMAX model introduces zeros (see the polynomial C in the numerator of definition [2.93]), but it conserves the same poles as those of the useful filter. In the BJ model, the disturbances filter is independent of the useful filter, with its own zeros and poles.

The polynomials specified in definitions [2.93] and [2.94] are expressed as:

$$\mathrm{C}\left(\mathrm{q}^{-1}\right) = 1 + c_1 \mathrm{q}^{-1} + \cdots + c_{nc} \mathrm{q}^{-nc}; \qquad [2.95]$$

$$\mathrm{D}\left(\mathrm{q}^{-1}\right) = 1 + d_1 \mathrm{q}^{-1} + \cdots + d_{nd} \mathrm{q}^{-nd}. \qquad [2.96]$$

In this section, a CLOE method adapted to identification of the ARMAX model is described. In the following section, another method is described, this time associated with the BJ model.

What makes the ARMAX model harder to identify than the ARX model is the fact that the regressors vector now includes values of the white noise. In order to obtain the configuration of this vector, one starts from the ARMAX model equation:

$$\begin{aligned} & y[n]+a_1 y[n-1]+\cdots+a_{na} y[n-na] \\ & \quad = b_0 u[n-k]+\cdots+b_{nb} u[n-k-nb] \\ & \quad + e[n]+c_1 e[n-1]+\cdots+c_{nc} e[n-nc], \ \forall n \in \mathbb{N}, \end{aligned} \quad [2.97]$$

which implies that:

$$\begin{cases} \boldsymbol{\varphi}[n]=\left[-y[n-1] \ \cdots -y[n-na] \ \mid u[n-k] \cdots u[n-k-nb] \ \mid e[n-1] \cdots e[n-nc]\right]^T \\ \boldsymbol{\theta}=\left[a_1 \ a_2 \ \cdots \ a_{na} \ \mid \ b_0 \ b_1 \cdots b_{nb} \ \mid \ c_1 \ c_2 \ \cdots \ c_{nc}\right]^T \qquad \forall n \in \mathbb{N} \end{cases}. \quad [2.98]$$

The vector $\boldsymbol{\varphi}$ is therefore only partially measurable. As a result, the LSM cannot be applied directly, but is instead applied following a two-stage strategy, which constitutes the core of *extended LSM* (E-LSM). First, a method to estimate the white noise contained in the I/O measured data has to be identified. This estimator is actually an auxiliary identification model, simpler than the original one. In turn, this model usually has a larger number of parameters. In the case of the ARMAX model, its adjuvant is an ARX model whose two polynomials are denoted by $\mathrm{A}_{n\alpha}$ and $\mathrm{B}_{n\beta}$ (with degrees $n\alpha$ and $n\beta$, respectively, significantly bigger than $na+nb+nc$). This model can be identified using a method taken from the LSM class. Second, the noise estimator is employed to replace the non-measurable values in the regressors vector with estimated values. The new vector then enables the LSM to be applied once again, in order to determine the coefficients of the three initial polynomials. One has to outline that $\mathrm{A}_{n\alpha}$ and $\mathrm{B}_{n\beta}$ are auxiliary polynomials only; they cannot replace the polynomials A and B of the ARMAX model.

For the CL process, if ARMAX replaces ARX as identification model, then the CLOE method becomes more complex and is referred to as the *eXtended CLOE* (X-CLOE).

In order to obtain the X-CLOE method, the E-LSM has to be adapted to the framework of Figure 2.3. To make the explanation easier to understand, the following equations have been linked to the basic CLOE method and not to the W-CLOE method. After understanding the idea of the X-CLOE method, the reader can begin to look at combining it with W-CLOE.

One starts with the preliminary identification described in section 2.3.3 for the ARX model. Since the result of this identification does not need to be extremely accurate, a simplified strategy can be used, even if the model of the process is now ARMAX.

Thus, of five polynomials determined for the ARX model, only three are kept, more specifically the ones that define the RST regulator. (In this identification step, to increase the accuracy, one can even consider that the indices of the ARX model are $[n\alpha, n\beta]$). Next, the approximate values of command u are evaluated, following equation [2.48]. This makes it possible to obtain the command and the output of process on the measurement horizon. The I/O dataset $\hat{\mathcal{D}}_N = \{\hat{u}[n], y[n]\}_{n\in\overline{1,N}}$ is therefore available. This means the E-LSM can be applied (as an OLSI method) by playing with the ARX and ARMAX models. The main equations used in this approach are the following (for any discrete instant $n \in \mathbb{N}$):

$$\begin{aligned}\hat{e}[n] = y[n] + \hat{\alpha}_1 y[n-1] + \cdots + \hat{\alpha}_{n\alpha} y[n-n\alpha] \\ - \hat{\beta}_0 \hat{u}[n-k] - \cdots - \hat{\beta}_{n\beta} \hat{u}[n-k-n\beta]\end{aligned} \qquad [2.99]$$

(estimate the white noise using the ARX model);

$$\left[\hat{\varphi}[n] = \left[-y[n-1] \ \cdots \ -y[n-na] \ \middle| \ \hat{u}[n-k] \cdots \hat{u}[n-k-nb] \ \middle| \ \hat{e}[n-1] \cdots \hat{e}[n-nc]\right]^T \qquad [2.100]\right.$$

(build the regressors vector for the ARMAX model);

$$\hat{\boldsymbol{\theta}}_N = \left(\frac{1}{N}\sum_{n=1}^{N} \hat{\varphi}[n]\hat{\varphi}^T[n]\right)^{-1} \left(\frac{1}{N}\sum_{n=1}^{N} \hat{\varphi}[n] y[n]\right) \qquad [2.101]$$

(estimate the vector of unknown parameters for the ARMAX model).

In equation [2.99], the coefficients of the ARX model have been written naturally. These coefficients are usually different from the estimated coefficients [2.101].

For online identification, one has to update not only the three polynomials of ARMAX model, but also the two polynomials of the auxiliary ARX model. Without these last two polynomials, the white noise cannot be estimated and the ARMAX model therefore cannot be updated.

The procedure of Algorithm 2.4 follows the basic CLOE method in spirit. The reader might want to design more complex algorithms, by adding weights, as explained in section 2.4.3.

1) *Initialization*:

a) Choose the structural index of the ARMAX model: $n\theta = na + nb + nc \in \mathbb{N}^*$.

b) Choose the structural index of the auxiliary ARX model: $n\tau = n\alpha + n\beta \in \mathbb{N}^*$ (commonly, $n\tau \geq 3n\theta$).

c) Identify a model of the CL process based on the ARX[nα,nβ] model, using one of the offline techniques presented in the previous section. This operation results in a preliminary estimation of five polynomials: $\tilde{A}_0$, $\tilde{B}_0$, $\hat{R}_0$, $\hat{S}_0$ and $\hat{T}_0$ (the first polynomials being auxiliary). An estimation of the process pure delay (k) is also available. This allows constructing the initial vector of the estimated parameters $\tilde{\boldsymbol{\theta}}_0$ from the coefficients of the auxiliary polynomials $\tilde{A}_0$ and $\tilde{B}_0$. An estimation of the virtual exogenous disturbance, $\hat{\mathbf{w}} \subset \mathbb{R}^2$, is available as well.

d) Estimate the command on the measurement horizon using the coefficients of polynomial $\hat{S}_0$ and the second component of the estimated virtual component:

$$\hat{u}[n] = \frac{1}{\hat{s}_{0,0}}\left(\hat{s}_{0,1}\hat{u}[n-1] - \cdots - \hat{s}_{0,ns}\hat{u}[n-ns] + \hat{w}_2[n]\right), \ \forall n \in \overline{1,N}.$$

e) Estimate the white noise on the measurement horizon:

$$\begin{aligned}\hat{e}[n] = {} & y[n] + \hat{\alpha}_1 y[n-1] + \cdots + \hat{\alpha}_{n\alpha} y[n-n\alpha] \\ & - \hat{\beta}_0 \hat{u}[n-k] - \cdots - \hat{\beta}_{n\beta}\hat{u}[n-k-n\beta], \ \forall n \in \overline{1,N}.\end{aligned}$$

f) Construct the regressors vectors for the ARMAX model:

$$\begin{aligned}\hat{\varphi}[n] = \big[& -y[n-1] \ \cdots - y[n-na] \ \big| \\ & \big| \ \hat{u}[n-k] \ \cdots \ \hat{u}[n-k-nb] \ \big| \\ & \big| \ \hat{e}[n-1] \ \cdots \ \hat{e}[n-nc] \big]^T, \ \forall n \in \overline{1,N}.\end{aligned}$$

g) Estimate the initial ARMAX model:

$$\hat{\boldsymbol{\theta}}_0 = \left(\frac{1}{N}\sum_{n=1}^{N}\hat{\varphi}[n]\hat{\varphi}^T[n]\right)^{-1}\left(\frac{1}{N}\sum_{n=1}^{N}\hat{\varphi}[n]y[n]\right).$$

The initialization of the CL identification model is now expressed by the polynomials: $\hat{A}_0$, $\hat{B}_0$, $\hat{C}_0$, $\hat{R}_0$, $\hat{S}_0$ and $\hat{T}_0$ (the first three being determined by the vector $\hat{\boldsymbol{\theta}}_0$ above).

h) Choose an initialization for the main matrices of ARX model: $\tilde{\mathbf{P}}_0 = \alpha^2 \mathbf{I}_{n\tau}$, with a large enough value of α (a few hundreds at least).

i) Choose an initialization for the main matrices of ARMAX model: $\hat{\mathbf{P}}_0 = \alpha^2 \mathbf{I}_{n\theta}$.

j) Initialize the estimated auxiliary regressors vector of the (ARX model) using the offline acquired data:

$$\tilde{\varphi}[1] = \left[-y[N-1] \ \cdots \ -y[N-n\alpha] \,\middle|\, \hat{u}[N-k] \ \cdots \ \hat{u}[N-k-n\beta]\right]^T .$$

k) Initialize the estimated main regressors vector (ARMAX mode) using the offline acquired data: $\hat{\varphi}[1] \leftarrow \hat{\varphi}[N]$.

l) Set the starting iteration index: $n = 1$.

2) *For* $n \geq 1$:

2.1. Acquire the current reference and output: $\{y_r[n], y[n]\}$.

2.2. Estimate the current command: $\hat{u}[n] = \dfrac{\hat{T}_{n-1}(q^{-1})}{\hat{S}_{n-1}(q^{-1})} y_r[n] - \dfrac{\hat{R}_{n-1}(q^{-1})}{\hat{S}_{n-1}(q^{-1})} \hat{\varphi}^T[n]\hat{\theta}_{n-1}$.

2.3. Prepare the auxiliary regressors vector for the next iteration (ARX model):

$$\tilde{\varphi}[n+1] = [-y[n] \ \cdots \ -y[n-n\alpha+1] \mid \hat{u}[n-k+1] \ \cdots \ \hat{u}[n-n\beta-k+1]]^T .$$

2.4. Evaluate the *a priori* output error (using the current ARMAX model): $\hat{\varepsilon}_y^b[n] = y[n] - \hat{\varphi}^T[n]\hat{\theta}_{n-1}$.

2.5. Evaluate the auxiliary vector (ARX model): $\tilde{\xi}_n = \tilde{\mathbf{P}}_{n-1}\tilde{\varphi}[n]$.

2.6. Evaluate the multiplicity constant of the output errors (ARX model): $\tilde{\mu}_n = \dfrac{1}{1 + \tilde{\varphi}^T[n]\tilde{\xi}_n}$ (scalar).

2.7. Evaluate the *a posteriori* output error (ARX model): $\tilde{\varepsilon}_y^f[n] = \tilde{\mu}_n \hat{\varepsilon}_y^b[n]$.

2.8. Update the main matrix of the ARX model: $\tilde{\mathbf{P}}_n = \tilde{\mathbf{P}}_{n-1} - \tilde{\mu}_n \tilde{\xi}_n \tilde{\xi}_n^T$.

2.9. Update the parameters vector for the ARX model: $\tilde{\theta}_n = \tilde{\theta}_{n-1} + \tilde{\xi}_n \tilde{\varepsilon}_y^f[n]$.

2.10. Evaluate the current white noise with the updated ARX model:

$$\begin{aligned}\hat{e}[n] &= y[n] + \hat{\alpha}_{n,1} y[n-1] + \cdots + \hat{\alpha}_{n,n\alpha} y[n-n\alpha] \\ &\quad - \hat{\beta}_{n,0}\hat{u}[n-k] - \cdots - \hat{\beta}_{n,n\beta}\hat{u}[n-k-n\beta].\end{aligned}$$

2.11. Prepare the main regressors vector for the next iteration (ARMAX model):

$$\begin{aligned}\hat{\varphi}[n+1] = [-y[n] \ &\cdots \ -y[n-na+1] \mid \\ &\mid \hat{u}[n-k+1] \ \cdots \ \hat{u}[n-k-nb+1] \mid \\ &\mid \hat{e}[n] \ \cdots \ \hat{e}[n-nc+1]]^T .\end{aligned}$$

2.12. Evaluate the auxiliary vector (ARMAX model): $\hat{\xi}_n = \hat{\mathbf{P}}_{n-1}\hat{\varphi}[n]$.

2.13. Evaluate the multiplicity constant of the output errors (ARMAX model): $\hat{\mu}_n = \dfrac{1}{1+\hat{\varphi}^T[n]\hat{\xi}_n}$ (scalar).

2.14. Evaluate the *a posteriori* output error (ARMAX model): $\hat{\varepsilon}_y^f[n] = \hat{\mu}_n \varepsilon_y^b[n]$.

2.15. Update the main matrix of ARMAX model: $\hat{\mathbf{P}}_n = \hat{\mathbf{P}}_{n-1} - \hat{\mu}_n \hat{\xi}_n \hat{\xi}_n^T$.

2.16. Update the vector of the parameters of the ARMAX model:

$$\hat{\theta}_n = \hat{\theta}_{n-1} + \hat{\xi}_n \hat{\varepsilon}_y^f[n].$$

2.17. Update the RST regulator $\{\hat{R}_n, \hat{S}_n, \hat{T}_n\}$ by using the ARMAX model $\{\hat{A}_n, \hat{B}_n, \hat{C}_n\}$ (built from $\hat{\theta}_n$).

2.18. Move on to the next iteration: $n \leftarrow n+1$.

3) *Return the model of CL process* $\{\hat{A}_n, \hat{B}_n, \hat{C}_n, \hat{R}_n, \hat{S}_n, \hat{T}_n\}$ *at every instant* $n \geq 0$. *An estimation of the command* $\hat{u}$ *is also available.*

Algorithm 2.4. *Steps of basic X-CLOE procedure (for the ARMAX model)*

The procedure of Algorithm 2.4 emphasizes the idea of an extended identification method. It is applied twice: first, for the auxiliary model and second, for the main model.

In steps 2.3 and 2.10, it is recommended that the next regressors vectors be prepared using the offline measured data during the initialization step. Moreover, there are recurrence relations between the successive regressors vectors, which can easily be seen by the reader. These recurrences help in efficiently implementing the algorithm. The required I/O data, which is measured or estimated offline, are already contained in the initial vectors.

This algorithm can usually improve the performance of CLOE algorithms based on the ARX model. Nevertheless, the fact that two of the ARMAX model filters have the same poles is a limitation. If the industrial application has enough computing power, it is better to work with a BJ model, which separates the two filters. The CLOE method adapted to this model is described in the following section.

2.4.6. *Generalized CLOE method*

It is normal to think that the external source of the disturbances affecting the dynamics of a process has little to do with the process itself. It is true that the process can interfere with other dynamic entities of the same environment. These interference disturbances usually affect all of the boxes that produced them. In order to generalize the approach concerning exogenous disturbances, it is better to work with the BJ model, described by the following main equation:

$$y[n] = \underbrace{q^{-k} \frac{B(q^{-1})}{A(q^{-1})}}_{\text{useful filter}} u[n] + \underbrace{\frac{C(q^{-1})}{D(q^{-1})}}_{\text{disturbances filter}} e[n], \quad \forall n \in \mathbb{N}. \qquad [2.102]$$

Polynomials A and D of the model can either be coprime or not. If interferences are a part of the exogenous disturbance, then it is better to accept some common poles for the two filters.

To identify the BJ model in a CL, a different approach to the previous one (specific to the X-CLOE method) has to be taken, since the regressors vector from definitions [2.98] no longer corresponds to a linear vector of the parameters (and there is no other way to build it). This limitation becomes obvious when equation [2.102] is expressed in equivalent form below:

$$A(q^{-1})D(q^{-1})y[n] = q^{-k}B(q^{-1})D(q^{-1})u[n] + A(q^{-1})C(q^{-1})e[n], \quad \forall n \in \mathbb{N}. \qquad [2.103]$$

The products of polynomials make the use of LSM inappropriate, as it can only provide the coefficients of these products and not the coefficients of each one of the four polynomials. Practically speaking, model [2.103] is of the ARMAX type. Even when it is identified, as in the previous section, it is very unlikely that the estimated products $\widehat{AD}$, $\widehat{BD}$ and $\widehat{AC}$ show all of the common roots for properly isolating the four polynomials.

The identification method therefore has to be changed. It is well known that linear models like BJ are quite precisely identified using the *prediction error minimization method* (PEMM). A description of this method is found in [STE 05, BOR 13a] or [BOR 13b], for example. It is one of the most general and accurate linear (and even nonlinear) identification methods. However, the price paid for this accuracy is the high computational complexity required by the corresponding numerical procedure.

The PEMM is based on an exact (gradient based) optimization method, namely the Gauss–Newton method (also described in [STE 05, BOR 13a] or [BOR 13b]). In order to alleviate the reader's task, we shall present a quick reminder of the two methods (Gauss–Newton and PEMM). Both have the goal of minimizing a quadratic criterion expressed as:

$$\mathcal{V}(\boldsymbol{\theta}) = \sum_{n=1}^{N} \varepsilon^2[n,\boldsymbol{\theta}], \tag{2.104}$$

where ε is generally known as the *modeling error*, while $N \in \mathbb{N}^*$ is the size of measurement horizon. For example, if this error is:

$$\varepsilon[n,\boldsymbol{\theta}] = y[n] - y_{\mathcal{M}}[n,\boldsymbol{\theta}], \tag{2.105}$$

for a given identification model $y_{\mathcal{M}}$, with the unknown parameters $\boldsymbol{\theta} \in \mathbb{R}^{n\theta}$, associated with a process whose output is y, then the quadratic criterion [2.51] is obtained (the same that started the presentation of CLOE methods).

The expression of modeling error is highly dependent on the mathematical model that describes the behavior of a particular entity (an industrial process, for instance). However, if this model makes the error at least once derivable with respect to vector $\boldsymbol{\theta}$, then a nonlinear programming method can be used to minimize criterion [2.104]. In this case, an effective method can be derived from the Newton–Raphson procedure (described in section 2.4.3). It is known as the Gauss–Newton method and its specific iterations are the following (with variable advancing steps):

$$\begin{cases} \boldsymbol{\theta}_{n+1} = \boldsymbol{\theta}_n + \alpha_n \mathcal{V}_{\boldsymbol{\theta\theta}}^{-1}(\boldsymbol{\theta}_n)\mathcal{V}_{\boldsymbol{\theta}}(\boldsymbol{\theta}_n) \\ \alpha_{n+1} = \alpha_n - \dfrac{\mathcal{V}_{\boldsymbol{\theta}}^T(\boldsymbol{\theta}_{n+1})\mathcal{V}_{\boldsymbol{\theta\theta}}^{-1}(\boldsymbol{\theta}_n)\mathcal{V}_{\boldsymbol{\theta}}(\boldsymbol{\theta}_n)}{\mathcal{V}_{\boldsymbol{\theta}}^T(\boldsymbol{\theta}_n)\mathcal{V}_{\boldsymbol{\theta\theta}}^{-1}(\boldsymbol{\theta}_n)\mathcal{V}_{\boldsymbol{\theta\theta}}(\boldsymbol{\theta}_{n+1})\mathcal{V}_{\boldsymbol{\theta\theta}}^{-1}(\boldsymbol{\theta}_n)\mathcal{V}_{\boldsymbol{\theta}}(\boldsymbol{\theta}_n)} \end{cases}, \quad \forall n \geq 0, \tag{2.106}$$

where

$$\mathcal{V}_{\boldsymbol{\theta}}(\boldsymbol{\theta}) = 2\sum_{n=1}^{N} \boldsymbol{\varepsilon}_{\boldsymbol{\theta}}[n,\boldsymbol{\theta}]\varepsilon[n,\boldsymbol{\theta}] \tag{2.107}$$

(the gradient of criterion) ;

$$\mathcal{V}_{\boldsymbol{\theta\theta}}(\boldsymbol{\theta}) = 2\sum_{n=1}^{N} \boldsymbol{\varepsilon}_{\boldsymbol{\theta}}[n,\boldsymbol{\theta}]\boldsymbol{\varepsilon}_{\boldsymbol{\theta}}^T[n,\boldsymbol{\theta}] \tag{2.108}$$

(the approximate Hessian matrix of the criterion).

Definition [2.108] shows that the second derivative of the modeling error does not need to be computed in order to build the Hessian matrix, which is a considerable advantage of this method.

The PEMM starts with equations [2.106]–[2.108] and proposes a method of computing the error gradient for each of the considered identification models. Its name comes from the term *prediction error* associated with the modeling error. Looking at definition [2.105], $y_{\mathcal{M}}[n,\boldsymbol{\theta}]$ can be considered to be the *current predicted value* of the process output, using the parameters $\boldsymbol{\theta}$ of the identification model. Consequently, $\varepsilon[n,\boldsymbol{\theta}]$ is the corresponding prediction error. In the case of ARMAX models in CL, the prediction error is identical to the *a priori* output error [2.61]. This justifies the term "prediction", as, to produce a current estimation of the process output, the previous identification model is used. In other words, a *predicted* estimation of the output at the next instant is obtained using the current identification model. Prediction error is actually an estimation of the exogenous noise that disturbs the process.

In the case of ARMAX model, the prediction error is expressed as follows:

$$\varepsilon[n,\boldsymbol{\theta}] = \mathrm{A}\left(\mathrm{q}^{-1}\right)y[n] - \mathrm{q}^{-k}\mathrm{B}\left(\mathrm{q}^{-1}\right)u[n] + \left[1-\mathrm{C}\left(\mathrm{q}^{-1}\right)\right]\varepsilon[n,\boldsymbol{\theta}], \quad \forall n \geq 0. \quad [2.109]$$

Equation [2.109] can be seen as a recurrent recipe to compute the prediction error starting from a given initialization, since the polynomial $1-\mathrm{C}$ has no constant term (and therefore produces a delay of at least one step on the applied signal). Moreover, this relationship shows that the prediction error depends linearly on the parameters of the model (the coefficients of the three polynomials), making it indefinitely derivable.

In fact, equation [2.109] expresses the technique on which PEMM is founded: computing the prediction error and its gradient in a recursive manner.

Also, equation [2.109] shows that the partial derivatives that make up the gradient $\boldsymbol{\varepsilon}_{\boldsymbol{\theta}}$ can be recursively calculated as below:

$$\frac{\partial \varepsilon}{\partial a_i}[n,\boldsymbol{\theta}] = \mathrm{q}^{-i}y[n] + \left[1-\mathrm{C}\left(\mathrm{q}^{-1}\right)\right]\frac{\partial \varepsilon}{\partial a_i}[n,\boldsymbol{\theta}], \quad \forall i \in \overline{1,na}, \ \forall n \geq 0; \quad [2.110]$$

$$\frac{\partial \varepsilon}{\partial b_j}[n,\boldsymbol{\theta}] = -\mathrm{q}^{-k-j}u[n] + \left[1-\mathrm{C}\left(\mathrm{q}^{-1}\right)\right]\frac{\partial \varepsilon}{\partial b_j}[n,\boldsymbol{\theta}], \quad \forall j \in \overline{0,nb}, \ \forall n \geq 0; \quad [2.111]$$

$$\frac{\partial\varepsilon}{\partial c_l}[n,\boldsymbol{\theta}] = -q^{-l}\varepsilon[n,\boldsymbol{\theta}] + \left[1 - C\left(q^{-1}\right)\right]\frac{\partial\varepsilon}{\partial c_l}[n,\boldsymbol{\theta}], \quad \forall l \in \overline{1,nc}, \ \forall n \geq 0. \quad [2.112]$$

Let us move on to the BJ model. In this case, the prediction error is (see the equivalent expression [2.103]):

$$\begin{aligned}\varepsilon[n,\boldsymbol{\theta}] &= A\left(q^{-1}\right)D\left(q^{-1}\right)y[n] - q^{-k}B\left(q^{-1}\right)D\left(q^{-1}\right)u[n] \\ &\quad + \left[1 - A\left(q^{-1}\right)C\left(q^{-1}\right)\right]\varepsilon[n,\boldsymbol{\theta}], \ \forall n \geq 0.\end{aligned} \quad [2.113]$$

As in the case of ARMAX model, the polynomial $1 - AC$ has no constant term, allowing the prediction error to be recursively computed using relation [2.113].

The gradient is practically defined using the following recurrent equations (derived from [2.113]):

$$\begin{aligned}\frac{\partial\varepsilon}{\partial a_i}[n,\boldsymbol{\theta}] &= q^{-i}D\left(q^{-1}\right)y[n] + \left[1 - A\left(q^{-1}\right)C\left(q^{-1}\right)\right]\frac{\partial\varepsilon}{\partial a_i}[n,\boldsymbol{\theta}], \\ &\forall i \in \overline{1,na}, \ \forall n \geq 0;\end{aligned} \quad [2.114]$$

$$\begin{aligned}\frac{\partial\varepsilon}{\partial b_j}[n,\boldsymbol{\theta}] &= -q^{-k-j}D\left(q^{-1}\right)u[n] + \left[1 - A\left(q^{-1}\right)C\left(q^{-1}\right)\right]\frac{\partial\varepsilon}{\partial b_j}[n,\boldsymbol{\theta}], \\ &\forall m \in \overline{1,nd}, \ \forall n \geq 0;\end{aligned} \quad [2.115]$$

$$\begin{aligned}\frac{\partial\varepsilon}{\partial c_l}[n,\boldsymbol{\theta}] &= -q^{-l}A\left(q^{-1}\right)\varepsilon[n,\boldsymbol{\theta}] + \left[1 - A\left(q^{-1}\right)C\left(q^{-1}\right)\right]\frac{\partial\varepsilon}{\partial c_l}[n,\boldsymbol{\theta}], \\ &\forall l \in \overline{1,nc}, \ \forall n \geq 0;\end{aligned} \quad [2.116]$$

$$\begin{aligned}\frac{\partial\varepsilon}{\partial d_m}[n,\boldsymbol{\theta}] &= q^{-m}A\left(q^{-1}\right)y[n] - q^{-k-m}B\left(q^{-1}\right)u[n] + \left[1 - A\left(q^{-1}\right)C\left(q^{-1}\right)\right]\frac{\partial\varepsilon}{\partial d_m}[n,\boldsymbol{\theta}], \\ &\forall m \in \overline{1,nd}, \ \forall n \geq 0.\end{aligned} \quad [2.117]$$

In the framework of CLOE methods, the equations to exploit are [2.113]–[2.117]. They constitute the core of the *generalized CLOE* (G-CLOE) method. The corresponding procedure, described in Algorithm 2.5, is quite different from the procedures described in the CLOE class, as, this time, the PEMM is used instead of the LSM.

1) *Initialization*:

a) Set the structural index of the BJ model: $n\theta = na + nb + nc + nd \in \mathbb{N}^*$.

b) Identify a model of the CL process based on ARX[nα,nβ] model, using one of the offline techniques presented in the previous section. This operation results in a preliminary estimation of five polynomials: $\hat{A}_0$, $\hat{B}_0$, $\hat{R}_0$, $\hat{S}_0$ and $\hat{T}_0$. Estimations of the process pure delay (k) and of the virtual exogenous disturbance ($\hat{\mathbf{w}} \subset \mathbb{R}^2$) are available as well.

c) Set the other two polynomials of the BJ model. Thus, $\hat{C}_0 = 1$ (with nc coefficients equal to zero), while $\hat{D}_0$ is built using some or all of the coefficients taken from the $\hat{A}_0$ polynomial, depending on which one is bigger between na and nd. If $nd < na$, the nd first coefficients of $\hat{A}_0$ are preserved. Otherwise, all coefficients of $\hat{A}_0$ are kept, and null values are added for the remaining coefficients. (Obviously, ARX is a particular type of BJ model).

d) Build the initial vector of estimated parameters, $\hat{\boldsymbol{\theta}}_0 \in \mathbb{R}^{n\theta}$, from the coefficients of polynomials: $\hat{A}_0 - 1$, $\hat{B}_0$, $\hat{C}_0 - 1$ and $\hat{D}_0 - 1$.

e) Build the initial parameters vector of the equivalent ARMAX model, $\tilde{\boldsymbol{\theta}}_0 \in \mathbb{R}^{2na+nb+nc+2nd}$, from the coefficients of polynomials: $\hat{A}_0\hat{D}_0 - 1$, $\hat{B}_0\hat{D}_0$ and $\hat{A}_0\hat{C}_0 - 1$.

f) Estimate the command on the offline measurement horizon, using the coefficients of polynomial $\hat{S}_0$ and the second component of the estimated virtual disturbance:

$$\hat{u}[n] = \frac{1}{\hat{s}_{0,0}}\left(\hat{s}_{0,1}\hat{u}[n-1] - \cdots - \hat{s}_{0,ns}\hat{u}[n-ns] + \hat{w}_2[n]\right), \ \forall n \in \overline{1,N}.$$

g) Estimate the white noise on the offline measurement horizon, using the identified ARX model:

$$\hat{e}[n] = \hat{A}_0\left(q^{-1}\right)y[n] - q^{-k}\hat{B}_0\left(q^{-1}\right)\hat{u}[n] + \left[1 - \hat{A}_0\left(q^{-1}\right)\right]\hat{e}[n], \ \forall n \in \overline{1,N}.$$

h) Construct the initial regressors vector corresponding to the equivalent ARMAX model:

$$\begin{aligned}\tilde{\boldsymbol{\varphi}}[1] = [-y[N] \ \cdots - y[N-na-nd+1] \,|\\ |\ \hat{u}[N-k+1] \ \cdots \ \hat{u}[N-k-nb-nd+1] \,|\\ |\ \hat{e}[N] \ \cdots \ \hat{e}[N-na-nc+1]]^T.\end{aligned}$$

i) Initialize the white noise:

$$\hat{e}[0] \leftarrow \hat{e}[N],\ \hat{e}[-1] \leftarrow \hat{e}[N-1],\ \ldots,\ \hat{e}[-n\theta] \leftarrow \hat{e}[N-n\theta].$$

j) Set a precision threshold $\delta > 0$ to stop the iterative procedure corresponding to PEMM.

k) Set the starting iteration index: $n = 1$.

2) *For* $n \geq 1$:

2.1. Acquire the current reference and output: $\{y_r[n], y[n]\}$.

2.2. Estimate the current command using the equivalent ARMAX model:

$$\hat{u}[n] = \frac{\hat{T}_{n-1}(q^{-1})}{\hat{S}_{n-1}(q^{-1})} y_r[n] - \frac{\hat{R}_{n-1}(q^{-1})}{\hat{S}_{n-1}(q^{-1})} \tilde{\boldsymbol{\varphi}}^T[n]\tilde{\boldsymbol{\theta}}_{n-1}.$$

2.3. Initialize the PEMM procedure:

$$\varepsilon\left[0, \hat{\boldsymbol{\theta}}_{n-1}\right] = \hat{e}[n-1],\ \varepsilon\left[-1, \hat{\boldsymbol{\theta}}_{n-1}\right] = \hat{e}[n-2],$$

$$\varepsilon\left[-n\theta, \hat{\boldsymbol{\theta}}_{n-1}\right] = \hat{e}[n - n\theta - 1];$$

$$\boldsymbol{\varepsilon}_{\boldsymbol{\theta}}\left[0, \hat{\boldsymbol{\theta}}_{n-1}\right] = \boldsymbol{\varepsilon}_{\boldsymbol{\theta}}\left[-1, \hat{\boldsymbol{\theta}}_{n-1}\right] = \cdots = \boldsymbol{\varepsilon}_{\boldsymbol{\theta}}\left[-n\theta, \hat{\boldsymbol{\theta}}_{n-1}\right] = \mathbf{0}\ ;\ \hat{\boldsymbol{\theta}}^0 = \hat{\boldsymbol{\theta}}_{n-1}\ ;\ \alpha^0 = 1.$$

2.4. For $r \in \overline{1,n}$:

2.4.1. Evaluate the prediction error:

$$\varepsilon\left[r, \hat{\boldsymbol{\theta}}_{n-1}\right] = \hat{A}_{n-1}(q^{-1})\hat{D}_{n-1}(q^{-1})y[r] - q^{-k}\hat{B}_{n-1}(q^{-1})\hat{D}_{n-1}(q^{-1})\hat{u}[r] + \left[1 - \hat{A}_{n-1}(q^{-1})\hat{C}_{n-1}(q^{-1})\right]\varepsilon\left[r, \hat{\boldsymbol{\theta}}_{n-1}\right].$$

2.4.2. Evaluate each component of the prediction error gradient:

$$\frac{\partial \varepsilon}{\partial a_i}\left[r, \hat{\boldsymbol{\theta}}_{n-1}\right] = q^{-i}\hat{D}_{n-1}(q^{-1})y[r] + \left[1 - \hat{A}_{n-1}(q^{-1})\hat{C}_{n-1}(q^{-1})\right]\frac{\partial \varepsilon}{\partial a_i}\left[r, \hat{\boldsymbol{\theta}}_{n-1}\right],\quad \forall i \in \overline{1,na};$$

$$\frac{\partial \varepsilon}{\partial b_j}\left[r, \hat{\boldsymbol{\theta}}_{n-1}\right] = -q^{-k-j}\hat{D}_{n-1}(q^{-1})\hat{u}[r] + \left[1 - \hat{A}_{n-1}(q^{-1})\hat{C}_{n-1}(q^{-1})\right]\frac{\partial \varepsilon}{\partial b_j}\left[r, \hat{\boldsymbol{\theta}}_{n-1}\right],\quad \forall j \in \overline{0,nb};$$

$$\frac{\partial \varepsilon}{\partial c_l}\left[r, \hat{\boldsymbol{\theta}}_{n-1}\right] = -q^{-l}\hat{A}_{n-1}(q^{-1})\varepsilon\left[r, \hat{\boldsymbol{\theta}}_{n-1}\right] + \left[1 - \hat{A}_{n-1}(q^{-1})\hat{C}_{n-1}(q^{-1})\right]\frac{\partial \varepsilon}{\partial c_l}\left[r, \hat{\boldsymbol{\theta}}_{n-1}\right],\quad \forall l \in \overline{1,nc};$$

$$\frac{\partial \varepsilon}{\partial d_m}\left[r,\hat{\boldsymbol{\theta}}_{n-1}\right]=\mathrm{q}^{-m}\hat{\mathrm{A}}_{n-1}\left(\mathrm{q}^{-1}\right)y[r]-\mathrm{q}^{-k-m}\hat{\mathrm{B}}_{n-1}\left(\mathrm{q}^{-1}\right)\hat{u}[r]$$
$$+\left[1-\hat{\mathrm{A}}_{n-1}\left(\mathrm{q}^{-1}\right)\hat{\mathrm{C}}_{n-1}\left(\mathrm{q}^{-1}\right)\right]\frac{\partial \varepsilon}{\partial d_m}\left[r,\hat{\boldsymbol{\theta}}_{r-1}\right],\ \forall m\in\overline{1,nd}.$$

2.5. Evaluate the correction matrix: $\mathbf{R}_n^0=\sum_{r=1}^{n}\boldsymbol{\varepsilon}_{\boldsymbol{\theta}}\left[r,\hat{\boldsymbol{\theta}}_{n-1}\right]\boldsymbol{\varepsilon}_{\boldsymbol{\theta}}^T\left[r,\hat{\boldsymbol{\theta}}_{n-1}\right]$.

2.6. Invert the matrix $\mathbf{R}_n^0$ (by considering its symmetry).

2.7. Evaluate the correction vector: $\mathbf{r}_n^0=\sum_{r=1}^{n}\boldsymbol{\varepsilon}_{\boldsymbol{\theta}}\left[r,\hat{\boldsymbol{\theta}}_{n-1}\right]\varepsilon\left[r,\hat{\boldsymbol{\theta}}_{n-1}\right]$.

2.8. For $p\geq 1$:

2.8.1. Evaluate the auxiliary vector: $\xi_n^{p-1}=\left(\mathbf{R}_n^{p-1}\right)^{-1}\mathbf{r}_n^{p-1}$.

2.8.2. Update the parameters vector: $\hat{\boldsymbol{\theta}}^p=\hat{\boldsymbol{\theta}}^{p-1}+\alpha^{p-1}\hat{\xi}_n^{p-1}$.

2.8.3. If $\left\|\hat{\boldsymbol{\theta}}^p-\hat{\boldsymbol{\theta}}^{p-1}\right\|=\left|\alpha^{p-1}\right|\left\|\hat{\xi}_n^{p-1}\right\|<\delta$, stop the iterative process, as the estimation of parameters has reached the required level of accuracy. Go directly to step 2.9.

2.8.4. Extract polynomials $\hat{\mathrm{A}}^p$, $\hat{\mathrm{B}}^p$, $\hat{\mathrm{C}}^p$ and $\hat{\mathrm{D}}^p$ from vector $\hat{\boldsymbol{\theta}}^p$.

2.8.5. For $r\in\overline{1,n}$:

a) Evaluate the prediction error:

$$\varepsilon\left[r,\hat{\boldsymbol{\theta}}^p\right]=\hat{\mathrm{A}}^p\left(\mathrm{q}^{-1}\right)\hat{\mathrm{D}}^p\left(\mathrm{q}^{-1}\right)y[r]-\mathrm{q}^{-k}\hat{\mathrm{B}}^p\left(\mathrm{q}^{-1}\right)\hat{\mathrm{D}}^p\left(\mathrm{q}^{-1}\right)\hat{u}[r]$$
$$+\left[1-\hat{\mathrm{A}}^p\left(\mathrm{q}^{-1}\right)\hat{\mathrm{C}}^p\left(\mathrm{q}^{-1}\right)\right]\varepsilon\left[r,\hat{\boldsymbol{\theta}}^p\right].$$

b) Evaluate the components of the prediction error gradient:

$$\frac{\partial \varepsilon}{\partial a_i}\left[r,\hat{\boldsymbol{\theta}}^p\right]=\mathrm{q}^{-i}\hat{\mathrm{D}}^p\left(\mathrm{q}^{-1}\right)y[r]+\left[1-\hat{\mathrm{A}}^p\left(\mathrm{q}^{-1}\right)\hat{\mathrm{C}}^p\left(\mathrm{q}^{-1}\right)\right]\frac{\partial \varepsilon}{\partial a_i}\left[r,\hat{\boldsymbol{\theta}}^p\right],\qquad \forall i\in\overline{1,na};$$

$$\frac{\partial \varepsilon}{\partial b_j}\left[r,\hat{\boldsymbol{\theta}}^p\right]=-\mathrm{q}^{-k-j}\hat{\mathrm{D}}^p\left(\mathrm{q}^{-1}\right)\hat{u}[r]+\left[1-\hat{\mathrm{A}}^p\left(\mathrm{q}^{-1}\right)\hat{\mathrm{C}}^p\left(\mathrm{q}^{-1}\right)\right]\frac{\partial \varepsilon}{\partial b_j}\left[r,\hat{\boldsymbol{\theta}}^p\right],\ \forall j\in\overline{0,nb};$$

$$\frac{\partial \varepsilon}{\partial c_l}\left[r,\hat{\boldsymbol{\theta}}^p\right]=-\mathrm{q}^{-l}\hat{\mathrm{A}}^p\left(\mathrm{q}^{-1}\right)\varepsilon\left[r,\hat{\boldsymbol{\theta}}^p\right]+\left[1-\hat{\mathrm{A}}^p\left(\mathrm{q}^{-1}\right)\hat{\mathrm{C}}^p\left(\mathrm{q}^{-1}\right)\right]\frac{\partial \varepsilon}{\partial c_l}\left[r,\hat{\boldsymbol{\theta}}^p\right],\ \forall l\in\overline{1,nc};$$

$$\frac{\partial \varepsilon}{\partial d_m}\left[r,\hat{\boldsymbol{\theta}}^p\right]=\mathrm{q}^{-m}\hat{\mathrm{A}}^p\left(\mathrm{q}^{-1}\right)y[r]-\mathrm{q}^{-k-m}\hat{\mathrm{B}}^p\left(\mathrm{q}^{-1}\right)\hat{u}[r]$$
$$+\left[1-\hat{\mathrm{A}}^p\left(\mathrm{q}^{-1}\right)\hat{\mathrm{C}}^p\left(\mathrm{q}^{-1}\right)\right]\frac{\partial \varepsilon}{\partial d_m}\left[r,\hat{\boldsymbol{\theta}}^p\right],\ \forall m\in\overline{1,nd}.$$

2.8.6. Evaluate the correction matrix: $\mathbf{R}_n^p=\sum_{r=1}^{n}\boldsymbol{\varepsilon}_{\boldsymbol{\theta}}\left[r,\hat{\boldsymbol{\theta}}^p\right]\boldsymbol{\varepsilon}_{\boldsymbol{\theta}}^T\left[r,\hat{\boldsymbol{\theta}}^p\right]$.

2.8.7. Invert the matrix $\mathbf{R}_n^p$ (by considering its symmetry).

2.8.8. Evaluate the correction vector: $\mathbf{r}_n^p=\sum_{r=1}^{n}\boldsymbol{\varepsilon}_{\boldsymbol{\theta}}\left[r,\hat{\boldsymbol{\theta}}^p\right]\varepsilon\left[r,\hat{\boldsymbol{\theta}}^p\right]$.

2.8.9. Prepare the adaptive step for the next iteration:

$$\alpha^p=\alpha^{p-1}-\frac{\mathbf{r}_n^p\hat{\xi}_n^{p-1}}{\left(\hat{\xi}_n^{p-1}\right)^T\mathbf{R}_n^p\hat{\xi}_n^{p-1}}$$

2.9. Update the BJ model: $\hat{\boldsymbol{\theta}}_n=\hat{\boldsymbol{\theta}}^p$. The new polynomials $\hat{\mathrm{A}}_n$, $\hat{\mathrm{B}}_n$, $\hat{\mathrm{C}}_n$ and $\hat{\mathrm{D}}_n$ are extracted from $\hat{\boldsymbol{\theta}}_n$.

2.10. Store in memory the current value of white noise: $\hat{e}[n]=\varepsilon\left[n,\hat{\boldsymbol{\theta}}^p\right]$.

2.11. Prepare the regressors vector of ARMAX model for the next iteration:

$$\tilde{\boldsymbol{\varphi}}[n+1]=\left[-y[n]\ \cdots -y[n-na-nd+1]\right|$$
$$\left|\ \hat{u}[n-k+1]\ \cdots\ \hat{u}[n-k-nb-nd+1]\right|$$
$$\left|\ \hat{e}[n]\ \cdots\ \hat{e}[n-na-nc+1]\right]^T.$$

2.12. Prepare the parameters vector of ARMAX model for the next iteration. Thus, $\tilde{\boldsymbol{\theta}}_n\in\mathbb{R}^{2na+nb+nc+2nd}$ includes the coefficients of polynomials: $\hat{\mathrm{A}}_n\hat{\mathrm{D}}_n-1$, $\hat{\mathrm{B}}_n\hat{\mathrm{D}}_n$ and $\hat{\mathrm{A}}_n\hat{\mathrm{C}}_n-1$.

2.13. Update the RST regulator: $\left\{\hat{\mathrm{R}}_n,\hat{\mathrm{S}}_n,\hat{\mathrm{T}}_n\right\}$ using the BJ model $\left\{\hat{\mathrm{A}}_n,\hat{\mathrm{B}}_n,\hat{\mathrm{C}}_n,\hat{\mathrm{D}}_n\right\}$.

2.14. Move on to the next iteration: $n\leftarrow n+1$.

3) *Return the model of CL process* $\left\{\hat{\mathrm{A}}_n,\hat{\mathrm{B}}_n,\hat{\mathrm{C}}_n,\hat{\mathrm{D}}_n,\hat{\mathrm{R}}_n,\hat{\mathrm{S}}_n,\hat{\mathrm{T}}_n\right\}$ *at each instant* $n\geq 0$. *An estimation of the command* $\hat{u}$ *is also available.*

Algorithm 2.5. *Steps of G-CLOE procedure (for the BJ model)*

The G-CLOE procedure from Algorithm 2.5 can turn into an X-CLOE procedure (for an ARMAX model), but it is sensibly different from the one in Algorithm 2.4 (as the PEMM is used instead of the E-LSM).

A criticism can be made of the G-CLOE algorithm: the computational effort is quite high (especially for real-time applications). Obviously, all computations of step 2 have to be performed within a timeframe at most equal to the sampling period. In this step however, cycles 2.4 and 2.8.4 take an increasing amount of time, as all of the acquired data are scanned from the beginning up to the current instant ($r \in \overline{1,n}$). Thus, the procedure becomes increasingly slow. Since the old I/O data should not influence the current dynamics of the process too much, the employment of some weighting window to be applied to the data is recommended. In SI, two types of window are commonly used:

– the exponential window: $w_{e,n}[r] = \lambda^{n-r}$, $\forall r \in \overline{1,n}$, where $\lambda \in (0,1]$ is a preset forgetting factor (for example, $\lambda = 0.98$);

– the rectangular window: $w_{r,n}[r] = \begin{cases} 1, & r \in \overline{n-M+1,n} \\ 0, & r \in \overline{1,n-M} \end{cases}$, where $M \in \mathbb{N}^*$ is preset and controls its aperture.

The exponential window does not reduce computational time, because the index r continues to vary from 1 to n. The rectangular window, however, limits its variation from $n-M+1$ to n. Despite its apparent advantage, the rectangular window has the drawback of introducing a significant disturbance into the data (especially if it is small in aperture), which increases the risk of degrading the identification model. Fortunately, the user can control the window aperture, such that a good trade off between the length of internal cycles in step 2 and the accuracy of the identification model is found. A combination of both windows can also be considered.

Another operation that can be time consuming is the matrices inversion. In the framework of the G-CLOE algorithm, this operation is planned in steps 2.6 and 2.8.6 for matrices that are similar to the Hessian matrix [2.108]. These matrices have $n\theta \times n\theta$ elements, resulting in a computational effort whose duration is proportional to $n\theta^3$ if the inversion is performed through the classical Gaussian algorithm. To reduce this effort, it is suitable to exploit the matrices symmetry, while using an inversion algorithm based on the QR decomposing with Givens-like rotations [GOL 96, PRE 07]. The inversion duration can thus decrease to a value proportional to $n\theta^2$.

The G-CLOE method exhibits very good performance in a number of industrial applications, as long as the computational complexity is reasonable.

2.4.7. CLOE methods for systems with integrator

It is of common practice to include an integrator right after the reference of a CL system in order to ensure the highest level of tracking. The integrator can improve the output dynamics, allowing the reference to be faster and more accurately tracked. Moreover, the integrator can contribute to attenuation (and even removal) of static errors, usually caused by uncalibrated sensors. Its system function is:

$$\mathrm{H}_I\left(\mathrm{q}^{-1}\right)=\frac{1}{1-\mathrm{q}^{-1}} \qquad [2.118]$$

During the application of identification and control, if the integrator is known to be part of the CL configuration, then it is better to take it into consideration. In the framework of CLOE methods, the integrator [2.118] (placed in front of the T component within the RST regulator) modifies the dynamics of the CL system (described in I/O equation [2.12]), as follows:

$$y \equiv \underbrace{\frac{\mathrm{q}^{-k}\mathrm{B}\left(\mathrm{q}^{-1}\right)\mathrm{T}\left(\mathrm{q}^{-1}\right)}{\left(1-\mathrm{q}^{-1}\right)\mathrm{P}\left(\mathrm{q}^{-1}\right)}}_{\text{useful filter}} y_r + \underbrace{\frac{\mathrm{A}\left(\mathrm{q}^{-1}\right)\mathrm{S}\left(\mathrm{q}^{-1}\right)}{\mathrm{P}\left(\mathrm{q}^{-1}\right)}}_{\text{disturbances filter}} v . \qquad [2.119]$$

Equation [2.119] is equivalent to:

$$\left(1-\mathrm{q}^{-1}\right) y \equiv \frac{\mathrm{q}^{-k}\mathrm{B}\left(\mathrm{q}^{-1}\right)\mathrm{T}\left(\mathrm{q}^{-1}\right)}{\mathrm{P}\left(\mathrm{q}^{-1}\right)} y_r + \frac{\mathrm{A}\left(\mathrm{q}^{-1}\right)\mathrm{S}\left(\mathrm{q}^{-1}\right)}{\mathrm{P}\left(\mathrm{q}^{-1}\right)}\left(1-\mathrm{q}^{-1}\right) v . \qquad [2.120]$$

The effect of the integrator on the model dynamics in equation [2.120] is remarkable. Instead of working with the successive values of output and disturbance, in this case, differences between successive values of output and disturbance are employed. More specifically, $\Delta y[n]=\left(1-\mathrm{q}^{-1}\right) y[n]=y[n]-y[n-1]$ replaces $y[n]$, while $\Delta v[n]=\left(1-\mathrm{q}^{-1}\right) v[n]=v[n]-v[n-1]$ replaces $v[n]$ (where $n \in \mathbb{N}^*$). Such differences can be referred to as *increments*. This explains the role of the integrator in a diagram of automatic regulation: the disturbance increments (instead of the disturbance itself) are processed and transformed into output increments (which are thus more accurately modified) in order to follow the reference more closely. Actually, the integrator is able to reject disturbances with systematic constant values (usually produced in the case of sensors decalibration, which leads to systematic errors in the acquired I/O data).

Identity [2.120] can be seen from a different angle. The incremental operator $(1-q^{-1})$ can be associated with system functions instead of signals. If the presence of the integrator is ignored, then polynomials A and R are enforced to have a unit root each, because:

$$\begin{aligned}(1-q^{-1})P(q^{-1}) = &\left[(1-q^{-1})A(q^{-1})\right]S(q^{-1}) \\ &+ q^{-k}B(q^{-1})\left[(1-q^{-1})R(q^{-1})\right].\end{aligned} \qquad [2.121]$$

The CLOE method should identify these roots, which is quite difficult because of numerical errors. Therefore, it is better to consider the existence of the integrator, if this information is known in advance.

In a CL diagram with the integrator, the command provided by the RST regulator is of the following form:

$$u \equiv \frac{T(q^{-1})}{(1-q^{-1})S(q^{-1})}y_r - \frac{R(q^{-1})}{S(q^{-1})}y$$

$$\Leftrightarrow \quad (1-q^{-1})u \equiv \frac{T(q^{-1})}{S(q^{-1})}y_r - \frac{R(q^{-1})}{S(q^{-1})}(1-q^{-1})y$$

$$\Leftrightarrow \quad \Delta u \equiv \frac{T(q^{-1})}{S(q^{-1})}y_r - \frac{R(q^{-1})}{S(q^{-1})}\Delta y\,. \qquad [2.122]$$

In equation [2.122], one can see that the command increments are produced by output increments (for the same reference), as expected.

If the disturbance is associated with a filter resulting from ARX, ARMAX or BJ models, one can generally write that:

$$\Delta v \equiv (1-q^{-1})v \equiv \frac{C(q^{-1})}{D(q^{-1})}(1-q^{-1})e \equiv \frac{C(q^{-1})}{D(q^{-1})}\Delta e, \qquad [2.123]$$

which also transfers the incremental computation to the white noise (note that if e is a centered, Gaussian white noise, then the signal of increments $\Delta e \equiv \left(1-\mathrm{q}^{-1}\right)e$ is also a centered, Gaussian white noise).

As a result, the CLOE algorithms described above can be modified as follows:

– the measured output y is replaced by the increments $\Delta y \equiv \left(1-\mathrm{q}^{-1}\right)y$;

– the estimated command $\hat{u}$ is replaced by the increments $\Delta\hat{u} \equiv \left(1-\mathrm{q}^{-1}\right)\hat{u}$;

– the disturbance v is replaced by the increments $\Delta v \equiv \left(1-\mathrm{q}^{-1}\right)v$;

– the white noise e is replaced by the increments $\Delta e \equiv \left(1-\mathrm{q}^{-1}\right)e$.

(Note that the reference remains unchanged.)

The changes above affect the whole process of identification from the initial determination of the crude ARX model to the end. For example, equation [2.43] becomes:

$$\underbrace{\begin{bmatrix} \mathrm{T}\left(\mathrm{q}^{-1}\right) & -\mathrm{R}\left(\mathrm{q}^{-1}\right) \\ -\mathrm{q}^{-k}\mathrm{Q}\left(\mathrm{q}^{-1}\right) & \mathrm{P}\left(\mathrm{q}^{-1}\right) \end{bmatrix}}_{\mathbf{A}\left(\mathrm{q}^{-1}\right)} \underbrace{\begin{bmatrix} y_r \\ \Delta y \end{bmatrix}}_{\mathbf{y}} \equiv \underbrace{\begin{bmatrix} \mathrm{S}\left(\mathrm{q}^{-1}\right)\Delta u \\ \mathrm{A}\left(\mathrm{q}^{-1}\right)\mathrm{S}\left(\mathrm{q}^{-1}\right)\Delta v \end{bmatrix}}_{\mathbf{w}}. \quad [2.124]$$

Also, definitions from [2.58] to [2.62] are transformed into (for any $n \in \mathbb{N}$):

$$\hat{\boldsymbol{\varphi}}[n] = \left[-\Delta y[n-1] \quad \cdots \quad -\Delta y[n-na] \mid \Delta\hat{u}[n-k] \quad \cdots \quad \Delta\hat{u}[n-nb-k]\right]^T; \quad [2.125]$$

$$\Delta\hat{y}_b[n] = \hat{\boldsymbol{\varphi}}^T[n]\hat{\boldsymbol{\theta}}_{n-1}; \quad [2.126]$$

$$\Delta\hat{y}_f[n] = \hat{\boldsymbol{\varphi}}^T[n]\hat{\boldsymbol{\theta}}_n; \quad [2.127]$$

$$\Delta\varepsilon_y^b[n] = \Delta y[n] - \Delta\hat{y}_b[n] = \Delta y[n] - \hat{\boldsymbol{\varphi}}^T[n]\hat{\boldsymbol{\theta}}_{n-1}; \quad [2.128]$$

$$\Delta\varepsilon_y^f[n] = \Delta y[n] - \Delta\hat{y}_f[n] = \Delta y[n] - \hat{\boldsymbol{\varphi}}^T[n]\hat{\boldsymbol{\theta}}_n. \quad [2.129]$$

This new class of methods is called *integral CLOE* (I-CLOE).

The basic I-CLOE procedure is described in Algorithm 2.6. The reader can rewrite all of the other CLOE algorithms for the regulation structure with integrator, if needed.

1) *Initialization*:

a) Set the structural index of the model: $n\theta = na + nb \in \mathbb{N}^*$.

b) Identify a CL model of the process using one of the offline techniques suggested in the previous section. This operation results in a preliminary estimation of five polynomials: $\hat{A}_0$, $\hat{B}_0$, $\hat{R}_0$, $\hat{S}_0$ and $\hat{T}_0$. Estimations of the process pure delay (k) and of command increments ($\Delta\hat{u}$) are also available. This allows constructing the initial vector of the estimated parameters, $\hat{\boldsymbol{\theta}}_0$, from the coefficients of polynomials $\hat{A}_0$ and $\hat{B}_0$. Note that this initialization step is based on the incremental equation [2.124].

c) Choose an initialization for the main matrices: $\mathbf{P}_0 = \alpha^2 \mathbf{I}_{n\theta}$, with a large enough value of α (a few hundreds, at least).

d) Initialize the estimated regressors vector:

$$\hat{\boldsymbol{\varphi}}[1] = \left[-\Delta y[N-1] \; \cdots \; -\Delta y[N-na] \;\middle|\; \Delta\hat{u}[N-k] \; \cdots \; \Delta\hat{u}[N-nb-k] \right],$$

where N is the size of the offline measurement horizon.

e) Set the starting iteration index: $n = 1$.

2) *For* $n \geq 1$:

2.1. Acquire the current reference and output: $\{y_r[n], y[n]\}$.

2.2. Estimate the increment of the current command:

$$\Delta\hat{u}[n] = \frac{\hat{T}_{n-1}(q^{-1})}{\hat{S}_{n-1}(q^{-1})} y_r[n] - \frac{\hat{R}_{n-1}(q^{-1})}{\hat{S}_{n-1}(q^{-1})} \hat{\boldsymbol{\varphi}}^T[n] \hat{\boldsymbol{\theta}}_{n-1}.$$

2.3. Estimate the current command: $\hat{u}[n] = \hat{u}[n-1] + \Delta\hat{u}[n]$.

2.4. Evaluate the current increment of the output: $\Delta y[n] = y[n] - y[n-1]$.

2.5. Prepare the regressors vector for the next iteration:

$$\hat{\boldsymbol{\varphi}}[n+1] = [-\Delta y[n] \;\; \cdots \;\; -\Delta y[n-na+1] \;\mid\; \Delta\hat{u}[n-k+1] \;\; \cdots \;\; \Delta\hat{u}[n-nb-k+1]]^T.$$

(Take values from vector $\hat{\boldsymbol{\varphi}}[1]$, if necessary)

2.6. Evaluate the increment of the a priori output error:

$$\Delta\varepsilon_y^b[n] = \Delta y[n] - \hat{\boldsymbol{\varphi}}^T[n] \hat{\boldsymbol{\theta}}_{n-1}.$$

2.7. Evaluate the auxiliary vector: $\hat{\xi}_n = \mathbf{P}_{n-1}\hat{\varphi}[n]$.

2.8. Evaluate the multiplicity constant of the output error increments:

$$\hat{\mu}_n = \frac{1}{1+\hat{\varphi}^T[n]\hat{\xi}_n} \text{ (scalar).}$$

2.9. Evaluate the *a posteriori* output error increment: $\Delta\varepsilon_y^f[n] = \hat{\mu}_n \Delta\varepsilon_y^b[n]$.

2.10. Update the main matrix: $\mathbf{P}_n = \mathbf{P}_{n-1} - \hat{\mu}_n \hat{\xi}_n \hat{\xi}_n^T$.

2.11. Update the vector of parameters: $\hat{\boldsymbol{\theta}}_n = \hat{\boldsymbol{\theta}}_{n-1} + \hat{\xi}_n \varepsilon_y^f[n]$.

2.12. Update the RST regulator, $\{\hat{R}_n, \hat{S}_n, \hat{T}_n\}$, from model $\{\hat{A}_n, \hat{B}_n\}$ (which, in turn, is built from $\hat{\boldsymbol{\theta}}_n$).

2.13. Move on to the next iteration: $n \leftarrow n+1$.

3) *Return the model of CL process* $\{\hat{A}_n, \hat{B}_n, \hat{R}_n, \hat{S}_n, \hat{T}_n\}$ *at each instant* $n \geq 0$. *An estimation of the command* $\hat{u}$ *is also available.*

Algorithm 2.6. *Steps of basic I-CLOE procedure*

Combinations between various algorithms of CLOE type are possible. Excluding the G-CLOE, usually the WX-CLOE (weighted X-CLOE) algorithm provides high performance and deserves to be employed in industrial applications.

2.4.8. *On the validation of CLOE identified models*

Although the identification techniques from the previous sections lead to online (real time) algorithms, it is necessary to check the validity of the involved models from time to time. The validation tests of CL models are similar to those used in the OLSI [STE 05, BOR 13a, BOR 13b].

In order to test the validity of CL identification models, it is wise to set up a simulation with these models, when using an I/O dataset other than the one employed to perform identification.

Overall, there are two categories of validation tests [LAN 93, LAN 01, LAN 05a]:

1) tests based on the techniques of decorrelation;

2) tests based on pole proximity.

The principle behind the first category of tests is to verify whether some signals from the diagram in Figure 2.3 are autocorrelated/crosscorrelated or not. The standard test involves deciding whether these signals have the characteristics of a centered, Gaussian white noise or not. (Therefore, this is also known as *whitening test.*) If the test is passed, then the identification is considered to be valid (even with a certain degree of validity – see [STE 05, BOR 13a, BOR 13b], for example). Otherwise, the model is invalid.

Usually, the following signals are checked out through the whitening test:

– the *a priori* output error, ε_y^b; if the error is similar to a centered, Gaussian white noise, then the identification model succeeds in separating the useful information from the parasite one, of the I/O data; this also means that the model fits the process, i.e. it is valid;

– the *a priori* output error versus the *a priori* output (provided by the process); in this case, the estimated crosscorrelation of the two signals:

$$\hat{r}_{\varepsilon_y^b,\hat{y}_b}[p] = \frac{1}{n-p}\sum_{i=p+1}^{n} \varepsilon_y^b[i]\hat{y}_b[i-p], \quad \forall p \in \overline{0,\left\lfloor \frac{n}{3} \right\rfloor}, \qquad [2.130]$$

where $n \in \mathbb{N}^*$ is the current instant, has to be close to the autocorrelation of a centered, Gaussian white noise; this can be explained by the fact that, if the model (ARX, ARMAX, BJ) of the process is valid, then it correctly integrates the dynamics of the disturbances source. As a result, the output error is only weakly correlated with the disturbances, while the simulated output is a very effective model of these disturbances.

For the second class of validation tests, the following technique is adopted: beside the CL identification of the process, an OLSI method is first employed, using the reference and the output as measured I/O signals in this aim; next, the poles of the general CL model are compared to the poles of the OL model. If the two models have poles that are located in approximately the same positions of the complex plane, then the CL model is valid.

Practically speaking, one focuses on the polynomial P [2.7], which can be estimated in two ways. On the one hand, a CLOE method can be used to obtain the estimation of P, say $\hat{P}_{CL}$ (see definition [2.7]). On the other hand, since its degree is known, P can be seen as the denominator of the useful system within an OL identification model (of ARX, ARMAX or BJ type). The OL identified model can be denoted by $\hat{P}_{OL}$. The degree of polynomials $\hat{P}_{CL}$ and $\hat{P}_{OL}$ must imperatively be the same. This allows the placement in the complex plane of their roots (i.e. the useful poles of the two models) to be compared. If the map of poles shows that they are grouped into pairs geometrically close to each other, then the CL model is considered valid. If, on the contrary, there is at least one pair of poles that are located far away from each other, then the CL model is invalid.

One has to outline that the test based on pole placement is more powerful than the whitening test. As a result, the number of models that can pass it is smaller.

2.5. Application: identification of active suspension

The application briefly presented in this last section of chapter is involved in an interesting problem that affects the automotive industry: the identification and automatic control of active suspension. A solution to this problem is presented in the following section, and the afferent explanation is focused on identification in CL. More details regarding this application can be found in [LAN 05b] and [STE 10], for example.

Nowadays, many important manufacturers in the automotive industry see active suspension as a viable alternative to classical suspension (which does not benefit from automatic control through electronic devices). There are some advantages in using this solution, which encourages the industry to standardize it in the future. First, with active suspension the lifespan of the shock dampers and the afferent mechanical components can be increased. Second, the car onboard computer can monitor the state of the suspension system, and quickly detect any failures. Finally, the comfort of the vehicle passengers (or the safety of carried goods) can be increased, as the absorption of vibrations during travel is carried out more efficiently. Other possible advantages can also be found. However, there is one main drawback: currently, the presence of automatic control in suspension system may considerably increase the vehicle price.

Research that has been carried out in order to solve the issue of active suspension has made use of the installation illustrated in Figure 2.5.

Figure 2.5. *Experimental active suspension system hosted by the GIPSA Laboratory in Grenoble, France*

This installation belongs to the GIPSA Laboratory (located in Grenoble, France). Despite appearances, the dynamics of the system tied to this setup are rather complex, with some natural resonance frequencies. Figure 2.6 shows the operating structure of the suspension system.

It highlights two subsystems: main and auxiliary. The main subsystem is made up of the classical suspension device, whose input is the primary force (seen here as a disturbance) produced by the vehicle and whose output is the residual force of the elastomeric cone (which partially dampens the primary force). (The forces can be measured using accelerometers.) The auxiliary system is an extension of the classical suspension device made up of a controllable inertial actuator. Its input is an electrical value (tension/current) applied through the winding coil in order to produce an additional dampening force. This force constitutes the output of the auxiliary system and is opposed to residual force of the elastomeric cone. When the two forces (residual and controlled dampening) are composed, a second residual force results (of smaller amplitude). This force is the input of controller (of RST type, and is possibly expressed in *Youla–Kucera form* [YK]), as illustrated in Figure 2.6.

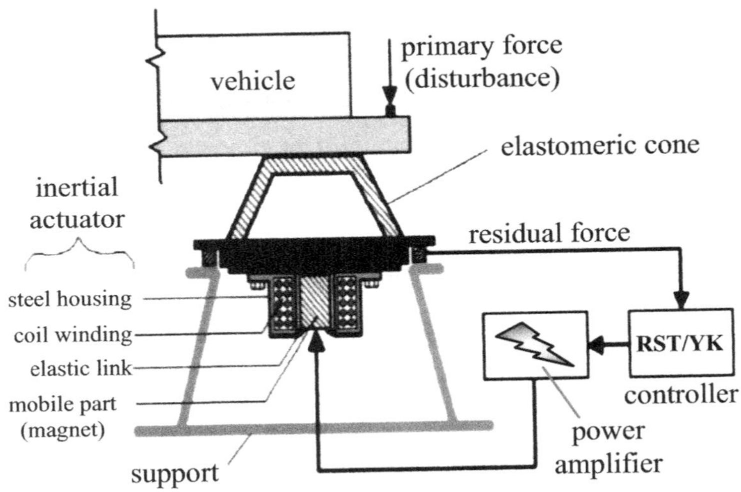

Figure 2.6. *Operating diagram of the experimental active suspension system hosted by the GIPSA Laboratory in Grenoble, France*

In this framework, the main problem is designing an RST regulator that can minimize the residual force, especially in the critical frequency bandwidths, where the installation exhibits resonances. Moreover, the regulator should be adaptive, given the nonlinear nature of installation. The CL auxiliary subsystem has to be previously identified in order to resolve this problem.

Before dealing with this problem, a global general offline identification of the active suspension system has been performed in the absence of any stochastic disturbances. The input is a PRS (converted to an electric value), with a large enough length (10,000 values). The sampling period is fixed at 800 Hz. Using the measured residual force, an ARX [12,14] model (with 12 poles and 14 zeros) is identified using a procedure belonging to the LSM class. This model made it possible to draw the smoothed spectrum of useful filter (in the absence of noise), as shown in Figure 2.7.

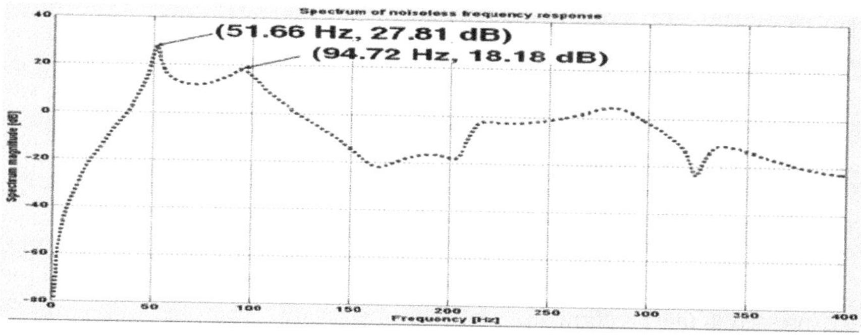

Figure 2.7. *Spectrum of the active suspension subsystem estimated in OL*

This curve emphasizes two (critical) resonance frequencies: one at 51.66 Hz (≅ 52 Hz) and another one at 94.72 Hz (≅ 95 Hz). Also, a nearly resonant behavior appears in the frequency band between 220 and 300 Hz. Nevertheless, considering the difference between the two highest resonance peaks (on the one hand) and the amplitude of spectrum in this band (on the other hand), one can admit that the dangerous resonance for the mechanical compounds of the system is the one corresponding to those peaks. The disturbances can amplify the resonance phenomenon if their spectrum exhibits important lines in the small subbands including the two critical frequencies. An issue tied to the general problem is, therefore, to dampen these resonances by optimal automatic control if possible. This involves the need to find appropriate CL identification models.

In order to obtain a coherent comparison between the results, two types of identification are tested: in OL and in CL. The models ARX, ARMAX and BJ are considered in both cases.

For OL identification, the following algorithms were implemented:

– adaptive extended LSM, recursive extended least squares (<u>R-ELS</u>) for the ARMAX model;

– *open loop output error* (<u>OLOE</u>) method for the ARX model;

– *extended OLOE* (<u>X-OLOE</u>) for the ARMAX model;

– PEMM for the BJ model (also known as *generalized least squares* [<u>G-LS</u>]).

These methods are extensively described in the scientific literature (see, for example, [SOD 89, LJU 99, DAU 04, STE 05, BOR 13a, BOR 13b]). Note that the only real difference between R-ELS and X-OLOE is that the model of the disturbance filter changes from C/A to $(C-A)/A$.

The CLSI methods employed here are:

– CLOE for the ARX model;

– X-CLOE for the ARMAX model;

– G-CLOE for the BJ model.

The RST regulator is implemented in two versions:

– classical (as presented in this chapter, see Figure 2.3);

– expressed in the YK form.

A comprehensive description of the latter can be found, for instance, in [BLO 06]. Figure 2.8 (actually derived from Figure 2.3) suggests the principle of RST regulators expressed in the YK form (RST-YK). At first sight, this parameterization is a seemingly useless complication, as a second reaction loop is inserted inside the existing loop, with an additional polynomial, L, to be determined. However, if the problem to solve is analyzed in detail, it appears that, in order to reduce the effect of disturbances in critical subbands, one has to control not only the useful filter, but also the other filter of the model, dealing with disturbances. The goal of the second reaction loop specifically is to control the disturbances filter.

NOTE.– In most publications, the RST-YK regulator is defined using another form of notation for the additional polynomial, namely Q. Since, in the framework of this chapter, Q refers to another polynomial, defined in [2.8], the original notation of this regulator has been changed to avoid any confusion.

The YK form of the RST regulator, illustrated in Figure 2.8, yields the conclusion that the CL system function is expressed by the following I/O relation (derived after some elementary algebraic manipulations):

$$y[n]=\frac{q^{-k}B(q^{-1})T(q^{-1})}{P(q^{-1})}y_r[n]+\frac{A(q^{-1})\left[S(q^{-1})-q^{-k}B(q^{-1})L(q^{-1})\right]}{P(q^{-1})}v[n],\ \forall n\in\mathbb{N},\quad [2.131]$$

where P is the same polynomial of classical RST (as defined in equation [2.7]).

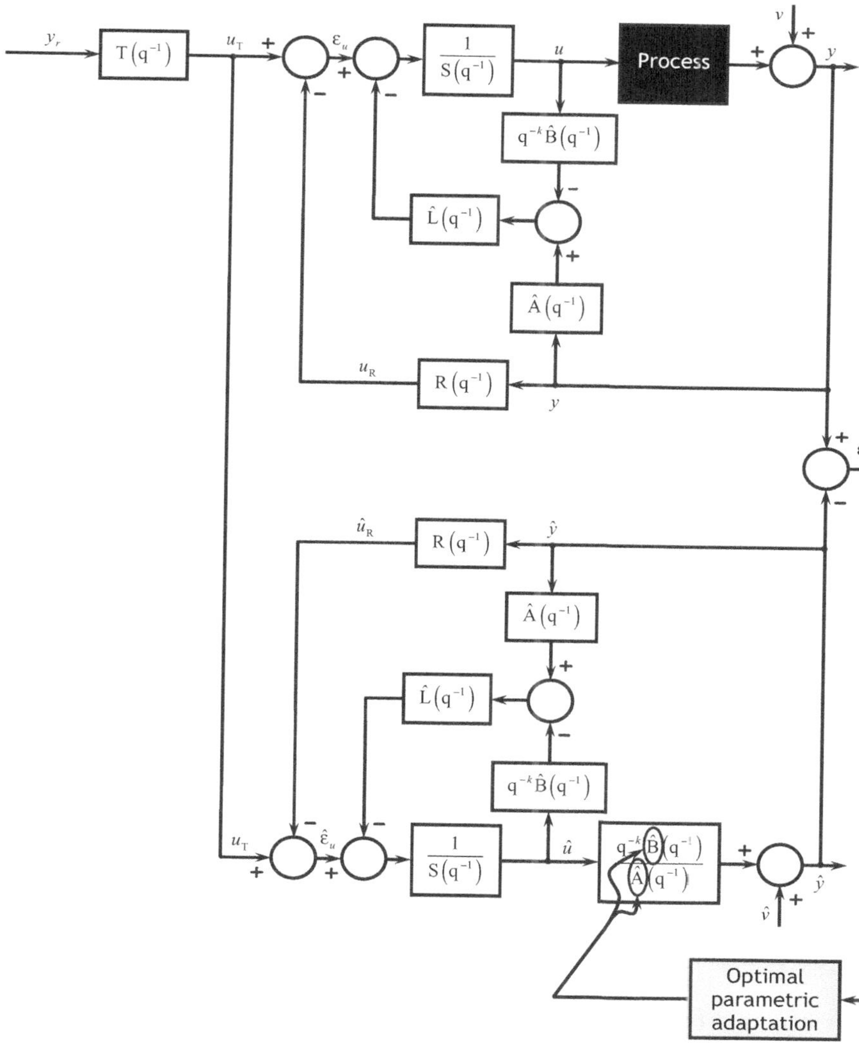

Figure 2.8. *Automatic control configuration with an RST-YK regulator*

Equation [2.131] now reveals the role of the supplementary polynomial L. One can consider the general case of the BJ model, which includes the ARX and ARMAX models as special cases. The disturbances filter is then [2.94], with the poles provided by polynomial D. In general, the frequency behavior of a linear system is more sensitive to the placement of its poles than to the placement of its zeros. It is therefore natural to remove the polynomial D in the expression of output [2.131], if possible. Fortunately, the polynomial L enables this operation. It suffices to consider the following factorization constraint:

$$\mathrm{S}\left(q^{-1}\right)-q^{-k}\mathrm{B}\left(q^{-1}\right)\mathrm{L}\left(q^{-1}\right)=\mathrm{L}_{\mathrm{D}}\left(q^{-1}\right)\mathrm{D}\left(q^{-1}\right), \qquad [2.132]$$

where L_{D} is a corresponding factor. If the equality [2.132] is verified by a couple $\{\mathrm{L},\mathrm{L}_{\mathrm{D}}\}$, then the I/O equation [2.131] becomes:

$$y[n]=\frac{q^{-k}\mathrm{B}\left(q^{-1}\right)\mathrm{T}\left(q^{-1}\right)}{\mathrm{P}\left(q^{-1}\right)}y_r[n]+\frac{\mathrm{A}\left(q^{-1}\right)\mathrm{L}_{\mathrm{D}}\left(q^{-1}\right)\mathrm{C}\left(q^{-1}\right)}{\mathrm{P}\left(q^{-1}\right)}e[n],\quad \forall n\in\mathbb{N}, \qquad [2.133]$$

which shows that the final filter of the white noise has the same poles as the CL system. Note that the presence of polynomial L in the regulation diagram only changes the contribution of the noise in the output and not the contribution of the reference, which is remarkable in itself.

In order to determine the couple $\{\mathrm{L},\mathrm{L}_{\mathrm{D}}\}$, equality [2.132] has to be employed. It can be seen as a Bézout-type polynomial identity. The degree of polynomial L can be varied (from a minimum value), in order to obtain several solutions (actually, a linear system has to be solved each time). Another approach involves first setting the polynomial L_{D} to a fixed expression (in order to obtain a certain frequency response of the white noise filter) and then determining the polynomial L by means of identity [2.132].

If the zeros of the disturbances filter need to be removed as well, then the following constraint is added to the RST regulator design method:

$$\mathrm{P}\left(q^{-1}\right)=\mathrm{A}\left(q^{-1}\right)\mathrm{S}\left(q^{-1}\right)-q^{-k}\mathrm{B}\left(q^{-1}\right)\mathrm{R}\left(q^{-1}\right)=\mathrm{P}_{\mathrm{C}}\left(q^{-1}\right)\mathrm{C}\left(q^{-1}\right), \qquad [2.134]$$

where P_{C} is the polynomial of the desired poles in CL. This involves transforming the I/O equation [2.133] into:

$$y[n]=\frac{q^{-k}\mathrm{B}\left(q^{-1}\right)\mathrm{T}\left(q^{-1}\right)}{\mathrm{P}_{\mathrm{C}}\left(q^{-1}\right)\mathrm{C}\left(q^{-1}\right)}y_r[n]+\frac{\mathrm{A}\left(q^{-1}\right)\mathrm{L}_{\mathrm{D}}\left(q^{-1}\right)}{\mathrm{P}_{\mathrm{C}}\left(q^{-1}\right)}e[n],\quad \forall n\in\mathbb{N}. \qquad [2.135]$$

The identity [2.134] makes the synthesis procedure of RST regulator more complicated and, moreover, if the there are any unstable roots of polynomial C, then the CL system becomes unstable (see the denominator of the first filter in equation [2.135]). Therefore, it is suitable to avoid the constraint [2.134], unless it is absolutely necessary and feasible.

Before concluding this section with some of the simulation results obtained using the above-mentioned identification methods, one has to outline that, in the case of the RST-YK regulator, the command is estimated through a different equation to [2.57], as follows:

$$\begin{aligned}\hat{u}[n] &= \frac{\mathrm{T}(q^{-1})}{\mathrm{S}(q^{-1})-q^{-k}\mathrm{B}(q^{-1})\mathrm{L}(q^{-1})}y_r[n]+\frac{\mathrm{A}(q^{-1})\mathrm{L}(q^{-1})-\mathrm{R}(q^{-1})}{\mathrm{S}(q^{-1})-q^{-k}\mathrm{B}(q^{-1})\mathrm{L}(q^{-1})}\hat{y}[n] \\ &= \frac{\mathrm{T}(q^{-1})}{\mathrm{L_D}(q^{-1})\mathrm{D}(q^{-1})}y_r[n]+\frac{\mathrm{A}(q^{-1})\mathrm{L_D}(q^{-1})\mathrm{D}(q^{-1})-\mathrm{R}(q^{-1})}{\mathrm{L_D}(q^{-1})\mathrm{D}(q^{-1})}\hat{y}[n],\ \forall n\in\mathbb{N}.\end{aligned} \quad [2.136]$$

Equation [2.136] shows that, on the one hand, the RST and RST-YK regulators are not equivalent and, on the other hand, the RST-YK regulator takes into consideration the poles of the disturbance filter directly.

The active suspension plant was tested for the three categories of disturbances, with the spectrum concentrated around frequencies of 75, 90 and 95 Hz, respectively. Because the spectrum of the third category of disturbances is in the zone of the second resonance peak of the characteristic shown in Figure 2.7, one expects that the rejection from the regulator is more difficult than for the other two categories of disturbances.

The reference was a PRS generated in the variation range of the residual force. Its spectrum is therefore slightly decreasing. If the regulator is enabled, then the residual force spectrum should be close to the reference spectrum. In the absence of the regulator, the reference is generated again, but this time within the variation range of the command signal, which directly stimulates the process. In this case, the resulting residual force does not have to follow it.

The goal of the first identification experiment was to select, among the proposed methods, the ones that produce the closest frequency characteristics to that shown in Figure 2.7 in the presence of disturbances. (Recall that the curve of Figure 2.7 was drawn in the absence of disturbances, and as such is "ideal".)

After conducting the corresponding simulations with the optimal identified models, the best models were identified to be G-LS in OL setting and G-CLOE in CL setting.

Figure 2.9 shows the spectrum of the best useful filters of the BJ models identified with the G-LS in OL setting (on the left column) and with the G-CLOE in CL setting (on the right column). In each image, the blue dotted curve represents the (pseudo)ideal spectrum of Figure 2.7, while the solid line curve (in red for the G-LS and in green for the G-CLOE) is the spectrum of the frequency response provided by the useful estimated filter ($\hat{B}/\hat{A}$). Likewise, each image corresponds to a disturbance (75, 90 and 95 Hz – from top down) and to a type of identification (in OL or in CL – from left to right). In order to compare the levels of performance of these filters, the distance in relation to the (pseudo)ideal spectrum is computed in the form of the standard deviation, denoted by λ . The smaller the value of λ, the higher the accuracy of the identified model.

These figures show that G-LS provides the most accurate models (compared with G-CLOE) – see the values of λ . This may seem surprising, but there is an explanation. For models in CL, the command is considered inaccessible during the data acquisition, while for OL models the command can be acquired. Practically speaking, since the command is always estimated and never measured in the framework of CLOE methods, it is normal to observe a slight reduction in quality for the BJ model.

The models can also be compared using the specific frequencies of disturbances. If the disturbance characteristic frequency is located between the two critical frequencies (notably at 75 Hz frequency), then the models are very close to the (pseudo)ideal one, which shows that the two filters of the BJ model conduct a proper separation of the two types of information contained in the data: useful and parasite. If this frequency approacles the critical frequencies (here, of 94.72 Hz), the accuracy of the identified model is degraded. Resonance makes the task of separating the two types of information more difficult. However, even more surprisingly, if the specific frequency of the disturbance is 95 Hz (practically stuck to the 94.72 frequency), then the precision of the models slightly increases. This can be explained by the fact that the identification model only mixes the two types of information in a tight subband surrounding a critical frequency, but clearly separates them outside of this subband. For the model, the disturbance at 95 Hz is part of the useful filter (see the bigger peak of the lined spectrum, compared to the dotted spectrum), but this filter remains unaffected by the other frequencies. The disturbance at 90 Hz, because of its closeness to the critical frequency of 94.72 Hz, increases the confusion between the two types of information, while this confusion is greatly attenuated for the disturbance at 95 Hz.

Following this first offline identification experiment, an initial BJ model can be proposed (for online methods) with $na = 10$, $nb = 12$, $nc = 1$, $nd = 12$, $k = 1$ in OL setting and with $na = 14$, $nb = 14$, $nc = 2$, $nd = 2$, $k = 1$ in CL setting.

The second identification experiment is conducted by using the adaptive versions of the two methods (G-LS and G-CLOE) in the same disturbance conditions. For G-LS, the classical RST regulator is used, while, for G-CLOE, the RST-YK regulator provides the command.

Figure 2.10 shows the variations of the residual force specific spectra: in the left column for the G-LS method and in the right column for the G-CLOE method. The small windows inserted into each spectral window represent the corresponding residual forces (which also reflect the instant at which the disturbances appeared: 1 s after the start of the variations). Each image corresponds to a disturbance (75, 90 and 95 Hz – from top down) and to an identification model (in OL setting or in CL setting – from left to right). On a spectral image, two variations are represented: without automatic controller (blue dotted line) and with automatic controller (solid red lines [G-LS] or solid green lines [G-CLOE]).

The G-LS method seemingly works in a similar way to the G-CLOE method, but the input to the active suspension subsystem is presumed accessible to the user, and therefore measurable (the reader can try to write out the afferent algorithm of identification and online control).

The comparison between the "inactive" (no controller) and "active" (with controller) spectra is carried out using two indicators:

– σ: the global standard deviation computed over the complete band of frequencies (from 0 to 400 Hz);

– σ_{nb}: the local standard deviation computed over a narrow band of frequencies around the critical frequency of 94.72 Hz (from 75 to 110 Hz).

The second standard deviation is considered to be the most important indicator, as the main goal is to dampen as much as possible the spectral lines around the critical frequency.

The images of Figure 2.10 first show that the "inactive" residual force actually contains three intrinsic resonances, as previously shown by the (pseudo)ideal spectrum of Figure 2.7 (visually compare it with the three peaks of the dotted curves in Figure 2.10). Moreover, the resonance around the frequency of 94.72 Hz appears to be the greatest of the three peaks. Second, it can be noted that the two controllers (RST and RST-YK) manage to quite significantly dampen not only the highest peak (94.72 Hz), but also the two other peaks (the first one just satisfactorily). Third, the active residual force spectrum seems to follow the reference spectrum in a satisfactory manner. However, one has to note that, usually, if the reference is a PRS (like in our case), then closely following it at the output of a CL process is a difficult task for any integrated automatic controller. This is the result of a well-known paradigm in Automatic Control: a system cannot be both indefinitely stable and fully adaptive at the same time. If the stability margin increases, then its adaptive capacity automatically decreases. However, if its reference tracking capability increases, then the risk of pushing it to the stability limits increases as well. In the case of PRS type references, the reduced tracking capacity of CL systems stems from the fact that the system has not finished its transient dynamics from one input value to the next. Practically speaking, the system is always in a transient state, leading it to states closer to instability than if it were in the steady state, where its dynamic behavior is far more stable.

Take a closer look at the performance levels in both regulation diagrams. The values of the global standard deviation σ show that the G-CLOE is slightly above the G-LS. In any case, the residual force contains low-frequency elements that are considerable. This can be partially explained by the inertia of the whole system, which is transferred to its output. Since the first peak is part of the low-frequency zone, it is normal that it would be less attenuated than the other peaks. However, attenuation of the last two peaks is clearly greater in both methods. This shows that the desired goal has been achieved.

The values of the local standard deviation, σ_{nb}, show that the G-CLOE, combined with the RST-YK, is better than the G-LS combined with the RST regulator, even though the difference is not too great.

In conclusion, as previously demonstrated within other practical applications, it can be said that OLSI techniques are not that far from CLSI techniques in terms of performance for automatic control. The main difference remains the feasibility. If the CL process does not allow measuring the command, then the CLSI techniques are the only ones that can be considered for effective implementation. On the contrary, due to the closeness of the performance levels within the two approaches, if the control signal is accessible to the user, there is a strong temptation to implement OLSI methods, which generally lead to less complex algorithms.

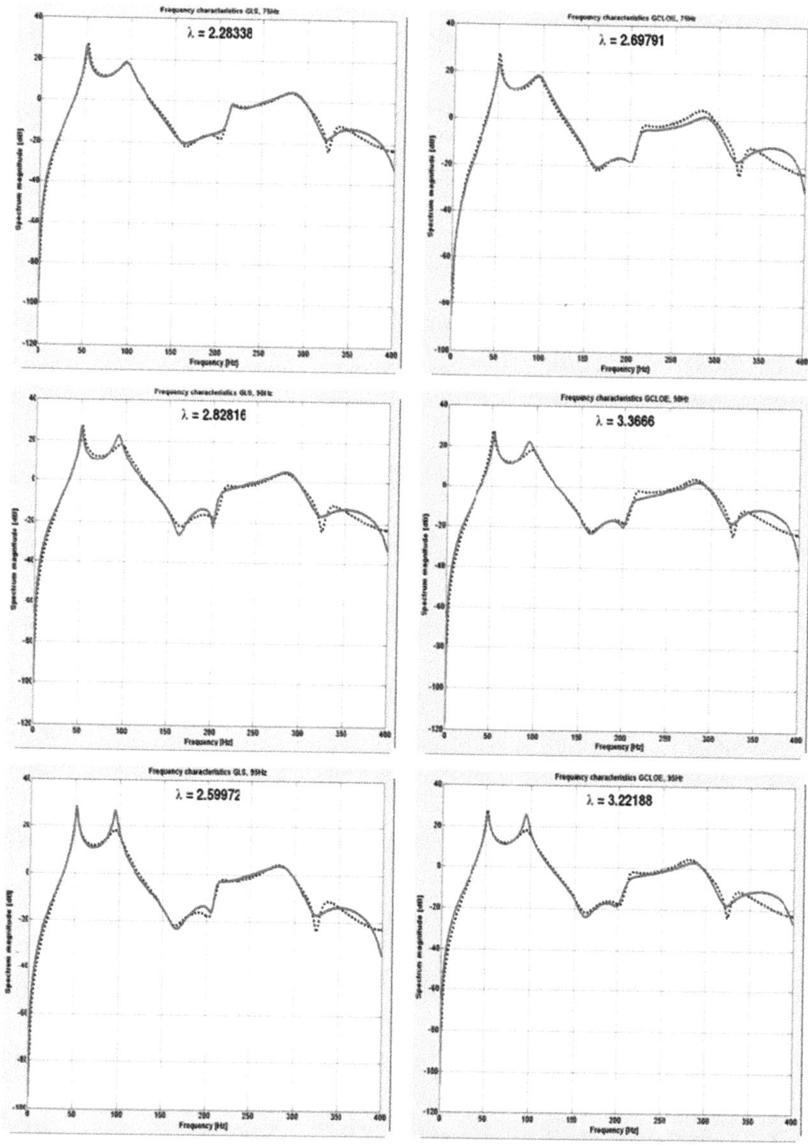

Figure 2.9. *Performance levels of the best identification methods for the active suspension subsystem*

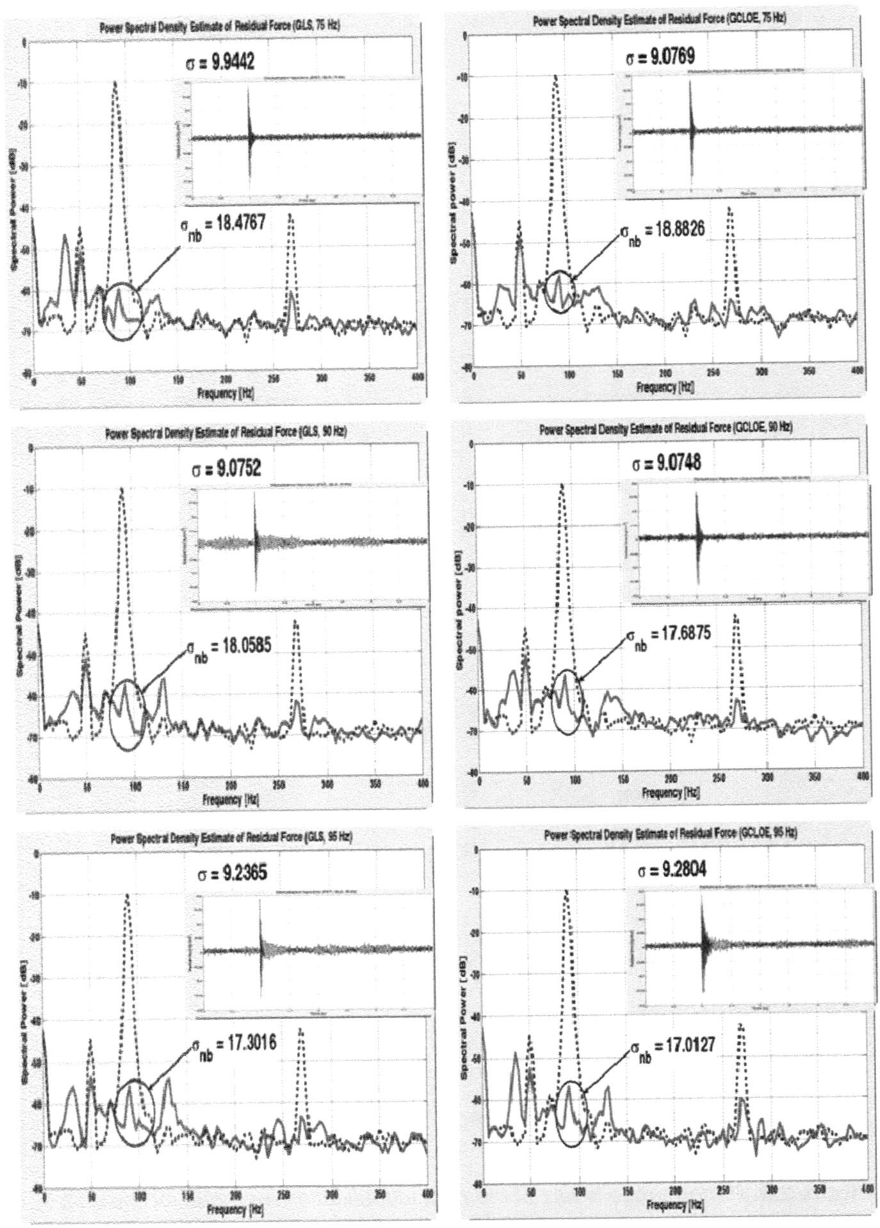

Figure 2.10. *Performance levels of the best solutions of RST and RST-YK control for the complete suspension system*

3

Digital Control Design Using Pole Placement

In this chapter, we shall present some numerical techniques for designing proportional-integral-derivative algorithm control, which is considered to be the classical control algorithm in automatic systems engineering, and the polynomial R (regulation), S (sensitivity) and T (tracking) (RST) control using a more flexible digital approach with two degrees of freedom, which is easy to integrate in real-time industrial applications. We have also developed a predictive control structure whose horizon is finite and unitary. The methods based on pole placement, which are the foundation for the development of these types of control algorithms, are used in the design of closed-loop systems, which ensure a high level of performance in driving industrial processes.

3.1. Digital proportional-integral-derivative algorithm control

For the design of proportional-integral-derivative PID algorithms in control systems, the required performances in the closed-loop system must be specified and the dynamic model of the process must be known (control and disturbance model).

Starting with this remark, the goal is to determine the structure and the parameters of the control algorithm.

This takes us back to the classic diagram of the continuous PID regulator, with filtering on the derivative component, presented in the literature [LAN 93, BIT 93, POP 00, POP 01, DAF 04].

$$G_{PID}(s) \stackrel{\Delta}{=} k\left[1+\frac{1}{T_i s}+\frac{T_d s}{\frac{T_d}{\lambda}s+1}\right]. \qquad [3.1]$$

The discrete version of relation [3.1], which uses a sampling period h, expressed in terms of the z transform is the following:

$$G_{PID}(z) = k\left[1 + \frac{\frac{h}{T_i}z}{z-1} + \frac{\frac{\lambda T_d}{T_d+\lambda h}(z-1)}{z - \frac{T_d}{T_d+\lambda h}}\right], \tag{3.2}$$

and in terms of z^{-1}

$$G_{PID}(z^{-1}) = k\left[1 + \frac{\frac{h}{T_i}}{1-z^{-1}} + \frac{\frac{\lambda T_d}{T_d+\lambda h}(1-z^{-1})}{1 - \frac{T_d}{T_d+\lambda h}z^{-1}}\right] = \frac{R(z^{-1})}{S(z^{-1})}, \tag{3.3}$$

where

$$\begin{cases} R(z^{-1}) \overset{\text{def}}{=} k(1-z^{-1})\left(1 - \frac{T_d}{T_d+\lambda h}z^{-1}\right) + \frac{kh}{T_i}\left(1 - \frac{T_d}{T_d+\lambda h}z^{-1}\right) \\ \qquad + \frac{k\lambda T_d}{T_d+\lambda h}(1-z^{-1})^2 \\ S(z^{-1}) \overset{\text{def}}{=} (1-z^{-1})\left(1 - \frac{T_d}{T_d+\lambda h}z^{-1}\right). \end{cases} \tag{3.4}$$

Polynomials R, S are therefore defined by the following expressions:

$$\begin{cases} R(Z^{-1}) = r_0 + r_1 Z^{-1} + r_2 Z^{-2} \\ S(Z^{-1}) = (1-Z^{-1})(1+S_1 Z^{-1}) \end{cases}. \tag{3.5}$$

Note that the PID controller is characterized by the four parameters, r_0, r_1, r_2 and s_1, and that the block diagram of the closed-loop system, shown in Figure 3.1, has its transfer function expressed as follows:

$$G_{RST}\left(z^{-1}\right) = \frac{\hat{b}\left(z^{-1}\right)R\left(z^{-1}\right)}{\hat{a}\left(z^{-1}\right)S\left(z^{-1}\right) + \hat{b}\left(z^{-1}\right)R\left(z^{-1}\right)}. \tag{3.6}$$

In the polynomial $S(z^{-1})$, the term $(1-z^{-1})$ ensures the effect of digital integration, while the term $(1+s_1 z^{-1})$ causes a digital filtering effect.

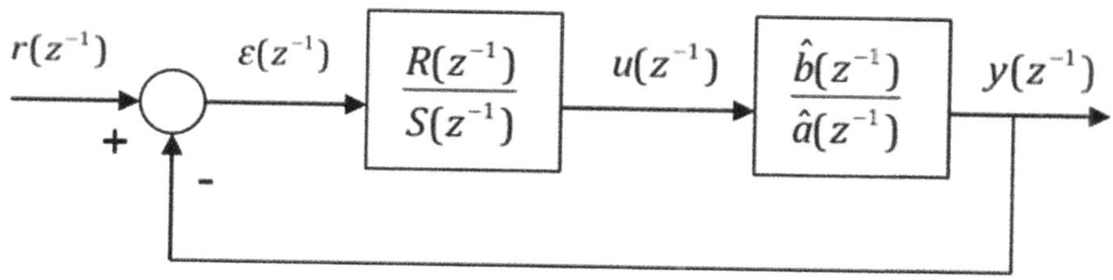

Figure 3.1. *PID-type control system in polynomial form*

The performance level of the system is specified by the desired poles for the system in the closed-loop system, the roots of the characteristic polynomial $P_i(z^{-1})$ (the denominator of [3.6]) being written as follows:

$$P_i(z^{-1}) \stackrel{\Delta}{=} \hat{a}(z^{-1})S(z^{-1}) + \hat{b}(z^{-1})R(z^{-1}). \qquad [3.7]$$

In practice, $P_i(z^{-1})$, which is recommended for the structure of the PID controller, is chosen in the following form:

$$P_i(z^{-1}) = 1 + p_1 z^{-1} + p_2 z^{-2}, \qquad [3.8]$$

which represents the discretized equivalent of a second-order continuous model.

As a result, calculation of parameters (R, S) of the PID controller assumes the resolution of the polynomial equation:

$$\hat{a}(z^{-1})(1-z^{-1})(1+s_1 z^{-1}) + \hat{b}(z^{-1})(r_0 + r_1 z^{-1} + r_2 z^{-2}) \equiv P_i(z^{-1}), \qquad [3.9]$$

where $\hat{a}(z^{-1})$, $\hat{b}(z^{-1})$ and $P_i(z^{-1})$ are known polynomials. The solution to this equation sets restrictions on the model of the process (of the second order at most with or without a time delay). It can be seen that the product $\hat{b}(z^{-1})R(z^{-1})$ defines the zeros in the closed-loop system; therefore, the PID controller introduces additional zeros through $R(z^{-1})$; the PID controller does not simplify the zeros of the process and can therefore be used for the regulation of processes whose models have unstable zeros; for the PID control system, performances are specified for regulation or tracking, and it is impossible to ensure independent dynamics for both tracking and regulation due to the same calculated PID control algorithm.

3.2. Digital polynomial RST control

In order to overturn the limitations of the PID algorithm, the structure in Figure 3.1 can be enriched using an additional tracking precompensating polynomial $T(z^{-1})$. The digital structure of the closed loop system is illustrated in Figure 3.2.

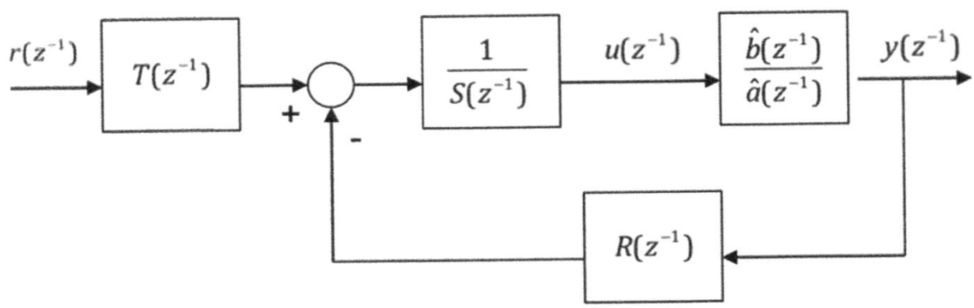

Figure 3.2. *RST control system*

The general algorithm of the digital control is expressed as follows:

$$u[k]=\frac{1}{s_0}\left[-\sum_{i=1}^{ns}s_i u[k-i]-\sum_{i=0}^{nr}r_i y[k-i]+\sum_{i=0}^{nt}t_i r[k-i]\right],\ \forall k\in\mathbb{N}. \qquad [3.10]$$

In this relation, $r[k]$ is the setpoint, $u[k]$ is the digital control, $y[k]$ is the system output and $\{s_i\}, i=0..ns$; $\{r_i\}, i=0..nr$; $\{t_i\}, i=0..nt$ are the parameters of the digital controller.

The number of coefficients provides the complexity and the memory of the algorithms, which are imposed depending on the size of the process and on the levels of performance required.

Starting with expression [3.10] above, it is possible to build the canonical (R (regulation), S (sensitivity) and T (tracking) (RST)) polynomial structure, which allows for the implementation of digital control systems in many industrial applications [BOR 93, BOR 97, GEN 97, LAN 97, POP 00, OGA 01, DAU 04].

The control algorithm can easily be expressed in an equivalent manner in z^{-1}:

$$u\left[z^{-1}\right]=\frac{T\left(z^{-1}\right)}{S\left(z^{-1}\right)}r\left[z^{-1}\right]-\frac{R\left(z^{-1}\right)}{S\left(z^{-1}\right)}y\left[z^{-1}\right], \qquad [3.11]$$

where

$$\begin{cases} R\left(z^{-1}\right)=r_0+r_1z^{-1}+...+r_{nr}z^{-nr} \\ S\left(z^{-1}\right)=s_0+s_1z^{-1}+...+s_{ns}z^{-ns} \\ T\left(z^{-1}\right)=t_0+t_1z^{-1}+...+t_{nt}z^{-nt}. \end{cases} \qquad [3.12]$$

The mathematical model of the controlled process is of the ARX type, expressed using the polynomials $\hat{a}(z^{-1})$ and $\hat{b}(z^{-1})$:

$$\begin{cases} \hat{b}\left(z^{-1}\right)=b_1z^{-1}+b_2z^{-2}+...+b_{nb}z^{-nb} \\ \hat{a}\left(z^{-1}\right)=1+a_1z^{-1}+...+a_{na}z^{-na} \end{cases}. \qquad [3.13]$$

The transfer function of the system represented in Figure 3.2 is the following:

$$G_{RST}\left(z^{-1}\right)=\frac{\hat{b}\left(z^{-1}\right)T\left(z^{-1}\right)}{\hat{a}\left(z^{-1}\right)S\left(z^{-1}\right)+\hat{b}\left(z^{-1}\right)R\left(z^{-1}\right)}. \qquad [3.14]$$

The configuration of the previous structure can still be improved using a trajectory generator in order to build a system that provides high levels of performance during tracking (setpoint change) and during regulation (disturbance rejection).

In this case, a reference trajectory $y^*[k]$ is built using the tracking model $\hat{b}_m(z^{-1})/\hat{a}_m(z^{-1})$ as shown in Figure 3.3, where:

$$\begin{cases} \hat{b}_m\left(z^{-1}\right)=b_{m1}z^{-1}+b_{m2}z^{-2}+...+b_{ms}z^{-ms} \\ \hat{a}_m\left(z^{-1}\right)=1+a_{m1}z^{-1}+...+a_{mr}z^{-mr}. \end{cases} \qquad [3.15]$$

The structure shown in Figure 3.3 specifies the dynamic behavior during tracking. This time the global transfer function is obtained using relation [3.14] (which expresses the effect of the regulation) by considering the additional action of the tracking model:

$$G^*_{RST}\left(z^{-1}\right)=\frac{\hat{b}_m\left(z^{-1}\right)}{\hat{a}_m\left(z^{-1}\right)}G_{RST}\left(z^{-1}\right)$$
$$=\frac{\hat{b}_m\left(z^{-1}\right)\hat{b}\left(z^{-1}\right)T\left(z^{-1}\right)}{\hat{a}_m\left(z^{-1}\right)\left[\hat{a}\left(z^{-1}\right)S\left(z^{-1}\right)+\hat{b}\left(z^{-1}\right)R\left(z^{-1}\right)\right]} \quad [3.16]$$

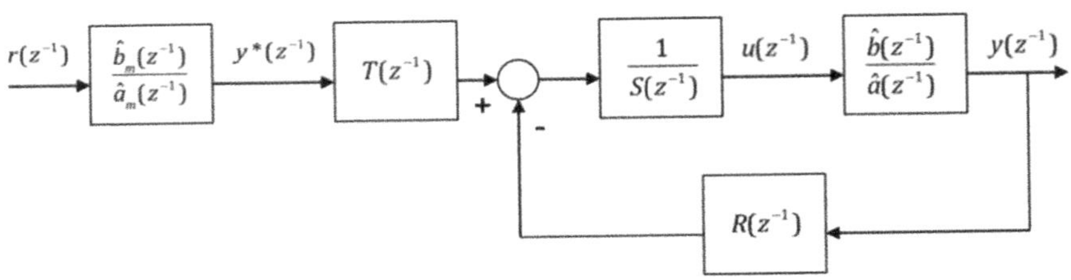

Figure 3.3. *RST structure with independent objectives in tracking and regulation*

3.3. RST control by pole placement

The placement of poles as shown is a strategy that allows a polynomial controller to be calculated without any restrictions on the degrees of the polynomials $\hat{a}(z^{-1})$ and $\hat{b}(z^{-1})$, and with no restriction on the delay of the process or on the zeros of the process (whether stable or unstable) [GOD 84, BOR 93, AST 97, LAN 97, DAU 04, POP 06].

The process is characterized by the irreducible model below:

$$G(z^{-1})=\frac{\hat{b}(z^{-1})}{\hat{a}(z^{-1})}, \quad [3.17]$$

where

$$\begin{cases}\hat{a}(z^{-1})=1+a_1z^{-1}+\ldots+a_{na}z^{-na}\\ \hat{b}(z^{-1})=b_1z^{-1}+\ldots+b_{nb}z^{-nb}.\end{cases} \quad [3.18]$$

In this case, the global transfer function of the closed-loop system is expressed as follows:

$$G_{RST}\left(z^{-1}\right)=\frac{T\left(z^{-1}\right)\hat{b}\left(z^{-1}\right)}{\hat{a}\left(z^{-1}\right)S\left(z^{-1}\right)+\hat{b}\left(z^{-1}\right)R\left(z^{-1}\right)}. \qquad [3.19]$$

3.3.1. *RST control for regulation dynamics*

Consider the polynomial that defines the level of performance regulation (disturbance rejection):

$$P_i\left(z^{-1}\right)\equiv 1+p_1 z^{-1}+p_2 z^{-2}+\ldots+p_n z^{-np}. \qquad [3.20]$$

From the previous relations, we can write that:

$$\hat{a}\left(z^{-1}\right)S\left(z^{-1}\right)+\hat{b}\left(z^{-1}\right)R\left(z^{-1}\right)\equiv P_i\left(z^{-1}\right). \qquad [3.21]$$

The polynomial equation [3.21] has a single solution in the following conditions:

$$\begin{cases} rank(\hat{a}(z^{-1}))=na \\ rank(\hat{b}(z^{-1}))=nb \\ rank(P(z^{-1}))=np\le na+nb. \\ rank(S(z^{-1}))=ns=nb \\ rank(R(z^{-1}))=nr=na \end{cases} \qquad [3.22]$$

To solve equation [3.21], we use the following matricial form, $Mx=p$

where:

$$\begin{cases} \boldsymbol{x}^{\mathrm{T}} \overset{\Delta}{=} \left[1\, s_1\, s_2 \ldots s_{nb}\, r_0\, r_1 \ldots r_{na-1}\right] \\ \boldsymbol{p}^{\mathrm{T}} \underset{def}{=} \left[1\, p_1\, p_2 \ldots p_{na+nb}\right]. \end{cases} \qquad [3.23]$$

M is the Sylvester matrix (of dimensions $(na + nb + 1) \times (na + nb + 1)$) associated with the polynomials $\hat{a}(z^{-1})$ and $\hat{b}(z^{-1})$ (for the polynomials $\hat{a}(z^{-1})$ and $\hat{b}(z^{-1})$ prime to one another, the Sylvester matrix is invertible).

Vector x that contains the unknown coefficients of polynomials $S(z^{-1})$ and $R(z^{-1})$ is then obtained simply by inversion of the matrix M:

$$x = M^{-1} p. \tag{3.24}$$

3.3.2. RST polynomial control for tracking dynamics (setpoint change)

In practice, the system output $y[k]$ needs to follow an imposed trajectory that is $y^*[k]$. This trajectory can be constructed using a trajectory generator, as shown in Figure 3.3.

The transfer function of this generator is defined by the dynamic model:

$$G_m\left(z^{-1}\right) \stackrel{\text{def}}{=} \frac{\hat{b}_m\left(z^{-1}\right)}{\hat{a}_m\left(z^{-1}\right)}, \tag{3.25}$$

where the polynomials $\hat{a}_m(z^{-1})$ and $\hat{b}_m(z^{-1})$ are introduced by the relations of [3.15]. In general, this transfer function is determined from the desired tracking performances, for example through a standard continuous second-order model.

Therefore, in practice, the discretized transfer function of the trajectory generator is represented with the imposed poles:

$$G_m\left(z^{-1}\right) = \frac{z^{-1}\left(b_{m0} + b_{m1} z^{-1}\right)}{1 + a_{m1} z^{-1} + a_{m2} z^{-2}}. \tag{3.26}$$

It is precisely why the controller must finally execute between $y^*[k]$ and the output $y[k]$, this transfer function. In the moment, this demand is not possible, as the zeros of the process (meaning the polynomial $b(z^{-1})$) are preserved.

The final objective will then be to reach the reference trajectory, which in this case is defined as follows:

$$y^*\left[z^{-1}\right] \stackrel{\text{def}}{=} \frac{\hat{b}_m\left(z^{-1}\right)}{\hat{a}_m\left(z^{-1}\right)} r\left[z^{-1}\right], \tag{3.27}$$

Polynomial $T(z^{-1})$ is chosen to ensure a unitary steady-state gain between $y^*[k]$ and $y[k]$ (to compensate the polynomial $P_i(z^{-1})$).

This leads to the following choice for polynomial $T(z^{-1})$:

$$T\left(z^{-1}\right)=GP_i\left(z^{-1}\right), \qquad [3.28]$$

where G is a constant gain, defined by:

$$G\triangleq\begin{cases}\dfrac{1}{\hat{b}(1)}, & \hat{b}(1)\neq 0\\ 1, & \hat{b}(1)=0\,.\end{cases} \qquad [3.29]$$

which leads to the following output:

$$y[z^{-1}]=G_{RST}(z^{-1})y^{*}[z^{-1}]=G\hat{b}(z^{-1})y^{*}[z^{-1}]. \qquad [3.30]$$

3.3.3. *RST control with independent objectives in tracking and regulation*

This approach results in the desired behavior of the system during tracking (setpoint change), independently from the behavior imposed during regulation (disturbance rejection), using the same control algorithm $u[z^{-1}]$.

Contrarily to the method of pole placement from section 3.3.2, this approach leads to the simplification of the stable zeros of the discrete model of the process, resulting in the desired performance during tracking and during regulation.

The structure of the closed loop system in question, as represented in Figure 3.3. is stable; once the transition is over, the output follows $y^*[z^{-1}]$ precisely.

The control strategy enables a digital (*RST*) controller to be calculated for stable or unstable processes, without any restriction on the degrees of the polynomials $\hat{a}(z^{-1})$ and $\hat{b}(z^{-1})$ of the discrete transfer function of the process. However, due to the simplification of the zeros, it can only be applied to models with stable zeros.

The output of the tracking model $\hat{b}_m(z^{-1})/\hat{a}_m(z^{-1})$ specifies the desired trajectory $y^*[z^{-1}]$ that the output of the closed-loop system must follow. As a result, $T(z^{-1})$ contributes to the calculation of the value of $y^*[z^{-1}]=y[z^{-1}]$.

The poles in a closed-loop system are defined by the desired characteristic polynomial $P_i\,(z^{-1})$.

Calculation of the polynomials $R(z^{-1})$, $S(z^{-1})$ and $T(z^{-1})$ takes place in two steps. First, using $R(z^{-1})$ and $S(z^{-1})$, the poles specified by $P_i(z^{-1})$ (regulation objective) are placed in a closed-loop system and the stable zeros of the discrete model of the process are simplified. In the second step, polynomial $T(z^{-1})$ is determined in order to find the reference trajectory $y^*[z^{-1}]$ at the output of the global control system.

3.3.3.1. *Performance in regulation (disturbance rejection)*

By eliminating the tracking part and the precompensator of the diagram in Figure 3.3, the transfer function in the closed-loop system with stable zeros is expressed as follows:

$$G_{RS}\left(z^{-1}\right)=\frac{\hat{b}\left(z^{-1}\right)}{\hat{a}\left(z^{-1}\right)S\left(z^{-1}\right)+\hat{b}\left(z^{-1}\right)R\left(z^{-1}\right)}. \quad [3.31]$$

Next, the following condition is imposed:

$$G_{RS}(z^{-1})=\frac{1}{P_i(z^{-1})}. \quad [3.32]$$

As polynomial $P_i(z^{-1})$ is specified *a priori*, this results in the following equation:

$$\hat{a}\left(z^{-1}\right)S\left(z^{-1}\right)+\hat{b}\left(z^{-1}\right)R\left(z^{-1}\right)=\hat{b}\left(z^{-1}\right)P_i(z^{-1}) \quad [3.33]$$

which allows $R(z^{-1})$ and $S(z^{-1})$ to be calculated.

Practically speaking, equation [3.33] demonstrates the manner in which the denominator of the transfer function $G_{RS}(z^{-1})$ must be factorized by $\hat{b}\left(z^{-1}\right)$. This operation leads to a natural factorization for $S(z^{-1})$:

$$S\left(z^{-1}\right)=\hat{b}\left(z^{-1}\right)S'\left(z^{-1}\right). \quad [3.34]$$

This results in a simpler expression of equation [3.33]:

$$\hat{a}\left(z^{-1}\right)S'\left(z^{-1}\right)+R\left(z^{-1}\right)\equiv P_i\left(z^{-1}\right). \quad [3.35]$$

Equation [3.35] has a single solution for:

$$\begin{cases} rank(\hat{a}(z^{-1}))=na \\ rank(P_i(z^{-1}))=np\leq na+ns' \\ rank(S'(z^{-1}))=ns' \\ rank(R(z^{-1}))=nr=na+ns' \end{cases}. \quad [3.36]$$

Equation [3.35] can be put into a matricial form $Mx = p$, where M is the Sylvester matrix, with dimensions of $(na + ns') \times (na + ns')$. Clearly:

$$\begin{bmatrix} x^T \overset{\Delta}{=} [1 \quad s_1 \quad s_2 \quad \dots \quad s_{ns'} \quad , \quad r_0 \quad r_1 \quad \dots \quad r_{na+ns'-1}] \\ p^T \overset{\Delta}{=} [1 \quad p_1 \quad p_2 \quad \dots \quad p_{na+ns'-1}] \end{bmatrix}. \qquad [3.37]$$

Here, the p_i are coefficients of the characteristic polynomial $P_i(z^{-1})$ from [3.35]. The solution for the polynomials from [3.35] is then obtained by inversion of matrix M:

$$x = M^{-1}p\,. \qquad [3.38]$$

3.3.3.2. *Tracking performance (setpoint change)*

Polynomial $T(z^{-1})$ is calculated by imposing the specified condition that the global structure in Figure 3.3 (obtained by adding the tacking part and the precompensator) has the same behavior as the tracking system:

$$G^*_{RST}\left(z^{-1}\right) \overset{\text{def}}{=} \frac{\hat{b}_m\left(z^{-1}\right)}{\hat{a}_m\left(z^{-1}\right)} T\left(z^{-1}\right) G_{RS}\left(z^{-1}\right) = \frac{\hat{b}_m\left(z^{-1}\right)}{\hat{a}_m\left(z^{-1}\right)}. \qquad [3.39]$$

We have already established that:

$$G_{RS}\left(z^{-1}\right) = \frac{1}{P_i\left(z^{-1}\right)} \qquad [3.40]$$

which leads to the identification of $T(z^{-1})$ with P_i: $T(z^{-1}) = P_i(z^{-1})$.

The regulator is then described by the following equation:

$$\begin{aligned} &S\left(z^{-1}\right)u\left[z^{-1}\right] + R\left(z^{-1}\right)y\left[z^{-1}\right] = P_i\left(z^{-1}\right)y^*\left[z^{-1}\right] \\ &\Leftrightarrow S\left(z^{-1}\right)u\left[z^{-1}\right] + R\left(z^{-1}\right)y^*\left[z^{-1}\right] = P_i\left(z^{-1}\right)y^*\left[z^{-1}\right]. \end{aligned} \qquad [3.41]$$

If we take into consideration the form of $S(z^{-1})$:

$$S\left(z^{-1}\right) = s_0 + s_1 z^{-1} + \dots + s_{ns} z^{-ns} = \hat{b}\left(z^{-1}\right) S'\left(z^{-1}\right), \qquad [3.42]$$

we can deduce that $s_0 = b_1$ and therefore:

$$S\left(z^{-1}\right)=b_1+z^{-1}S^*\left(z^{-1}\right). \qquad [3.43]$$

The control algorithm (equation of the controller) for implementation is then:

$$u[z^{-1}]=\frac{1}{b_1}(P_i(z^{-1})y[z^{-1}]-S'(z^{-1})u[z^{-1}]-R(z^{-1})y^*[z^{-1}]). \qquad [3.44]$$

3.4. Predictive RST control

This section presents the technique of predictive control, which is often used in advanced digital control systems. We will begin by presenting the key ideas behind its design and the purpose of this approach. Next, we will highlight the various algorithmic steps of predictive control. Finally, we describe the most representative methods ("dead beat") for predictions in the range of a sample period [FLA 94, MAC 01, POP 04].

The calculation of predictive control is characterized by the following steps:

1) at each instant kh (where h is the sampling period), a prediction of the evolution of the process is carried out for a sequence of future actions on the input;

2) the digital control sequence is determined by optimizing a criterion based on the future values; the calculation takes into account the difference between the predicted value and the setpoint as well as the variations of the input;

3) only the first control of the sequence is effectively applied and a new calculation begins, taking into account the evolution of the process in real time.

At an instant $t = kh$, the next output of the process is predicted over a time horizon from a mathematical model of the process dynamics. This depends on the past evolution of the process and the control scenario suggested for the next period.

Depending on various solutions, we choose the control sequence that takes the output of the process to the value of the setpoint in an "optimal" manner in relation to an objective defined beforehand. The calculated control is applied to the real process at the instant kh up to the instant $(k + 1)h$.

The advantage of this approach lies in the fact that predictive control is better than PID control for the delayed process when the setpoint changes are known or programmed ahead. Predictive control is considered to be robust for a rough approximation of the controlled process.

An important parameter of predictive control is the *horizon*, which has several representations, as follows:

– *the model horizon* (hM) is the length of time taken for the response of the process in an open loop to reach 99% of its final value;

– *the control horizon* (hC) is the length of the application time of the successive actions on the input of the process needed in order to bring the output to the desired value;

– *the prediction horizon* (hP) is the length of time over which the output of the process is predicted and used in the recurrent calculation of the control in order to anticipate the setpoint.

We can consider the particular case corresponding to a predictive control for a horizon reduced to a single step. The model for the prediction of the output is given by identification from the data measured at the input and the output of the process. The calculation of the predictive control algorithm therefore involves deducing value $u[k]$, which simply ensures that the next output $y[k+1]$ is equal to the current setpoint $r[k]$. In this case, a prediction horizon is then equal to only one sampling period ($hP = h$).

In most situations a unitary horizon is sufficient, and we shall show that in such a case, predictive control is simple to determine and easy to implement.

3.4.1. *Finite horizon predictive control*

The algorithms of predictive control are based on the prediction of the output of the process. To calculate the predicted value, a model of the process is used. The easiest one to obtain is the model built from a step response, for example.

Let us consider a step response for a finite series of values a_i ($a_1, a_2,\ldots,a_{hM}$) taken at the instants (1, 2, 3,..., hM), which represents a convolution model. This model can be used to calculate the response of the system following a variation at the input.

We assume $y[0] = 0$ and that the input is a series of variations of the input (Δu_0 at $k = 0$ and Δu_i at $k = i$). The output of the system is then given by the following model:

$$y[k] = \sum_{i=1}^{hM} a_i \Delta u_{k-1}, \quad \forall k \in \mathbb{N}, \qquad [3.45]$$

where hM is the model horizon.

The weighted model can be defined as the derivative of the step response. In this case, the output of the system is described by a similar model as follows:

$$y[k]=\sum_{i=1}^{hM} h_i \Delta u_{k-1}, \quad \forall k \in \mathbb{N}, \qquad [3.46]$$

where $h_i = a_i - a_{i-1}, i \in \overline{1,hM}$. For a given possible delay d, the first coefficients are equal to zero: $h_i = 0$ ($i \in \overline{1,d}$).

The prediction of the output on the horizon hC ($y_p[1]$, $y_p[2]$,...,y_p $[hC]$), for a sequence of hM actions (a_1, a_2,...,a_{hM}), based on convolution response model, can be written in the matricial form:

$$\underbrace{\begin{bmatrix} y_p[k+1] \\ y_p[k+2] \\ \vdots \\ y_p[k+hC] \end{bmatrix}}_{Y_p[k]} \stackrel{def}{=} \underbrace{\begin{bmatrix} a_1 & 0 & 0 & \cdots & 0 \\ a_2 & a_1 & 0 & \cdots & 0 \\ \vdots & \vdots & \ddots & \vdots & \vdots \\ a_{hC} & a_{hC-1} & \cdots & \cdots & a_{hC-hM-1} \end{bmatrix}}_{A} \cdot \underbrace{\begin{bmatrix} \Delta u_k \\ \Delta u_{k+1} \\ \vdots \\ \Delta u_{k+hM-1} \end{bmatrix}}_{\Delta U[k]} . \qquad [3.47]$$

If the control horizon hC is greater than the model horizon hM, we state that $a_i = a_{hM}$ for $i > hM$ and we write:

$$Y_p[k] = A \Delta U[k], (\forall) k \in \mathbb{N}. \qquad [3.48]$$

The matrix A is obtained from the model of the step response. Relation [3.48] allows us to calculate the control variation $\Delta U[k]$. The control objective is to force the system output to follow the reference trajectory $r[k + 1]$, $r[k + 2]$,..., $r[k + hC]$. To do this, the minimization of a quadratic criterion is imposed:

$$J[k] = \sum_{i=1}^{hC} (r[k+i] - y_p[k+i])^2 + \lambda \sum_{i=1}^{hM} \Delta u^2[k+i]. \qquad [3.49]$$

The vector $\Delta U[k]$ that minimizes the criterion $J[k]$ by the method of least squares is given in the following relation:

$$\Delta U[k] = (A^T A + \lambda^2 I)^{-1}(R[k] - Y_p[k]), \qquad [3.50]$$

where

$$R^{\mathrm{T}}[k] \overset{def}{=} \left[r[k+1] \ldots r[k+hC] \right], \quad \forall k \in \mathbb{N}. \tag{3.51}$$

Only the first element of $\Delta U[k]$ is used to update the control:

$$u[k] = u[k-1] + \Delta u[k], \quad \forall k \in \mathbb{N}, \tag{3.52}$$

which guarantees the desired evolution of the output $y[k]$.

3.4.2. *Predictive control with unitary horizon*

As stated above, a predictive control developed over a unitary horizon is sufficient in many cases. In this case, the prediction horizon is equal to a sampling period, while the model horizon hM and control horizon hC are also reduced to a single sampling period.

The process model, which specifies the predicted output, is assumed to have been known through an identification operation.

In Figure 3.4, we present the temporal evolution of the process output, of the desired output and the corresponding predictive control, for a unitary prediction horizon. These representations show the continual evolution of the corresponding values between the two successive sampling instants.

The structure of the closed-loop system is specified in Figure 3.2. We can consider that the model of the process is of an ARX type system, whose transfer function is the following:

$$G\left(z^{-1}\right) = \frac{\hat{b}\left(z^{-1}\right)}{\hat{a}\left(z^{-1}\right)}, \tag{3.53}$$

where

$$\begin{bmatrix} \hat{a}(z^{-1}) = 1 + a_1 z^{-1} + a_2 z^{-2} + \ldots + a_{na} z^{-na} \\ \hat{b}(z^{-1}) = b_1 z^{-1} + b_2 z^{-2} + \ldots + b_{nb} z^{-nb} \end{bmatrix}. \tag{3.54}$$

Relations [3.53] and [3.54] allow $y[k]$ to be expressed as follows:

$$y[k] = -\sum_{i=1}^{na} a_i y[k-i] + \sum_{i=1}^{nb} b_i u[k-i], \quad \forall k \in \mathbb{N}. \qquad [3.55]$$

The predictor of the process is determined simply by the progress of a sampling period:

$$y[k+1] = \sum_{i=1}^{na} a_i y[k-i+1] + \sum_{i=1}^{nb} b_i u[k-i+1], \quad \forall k \in \mathbb{N}. \qquad [3.56]$$

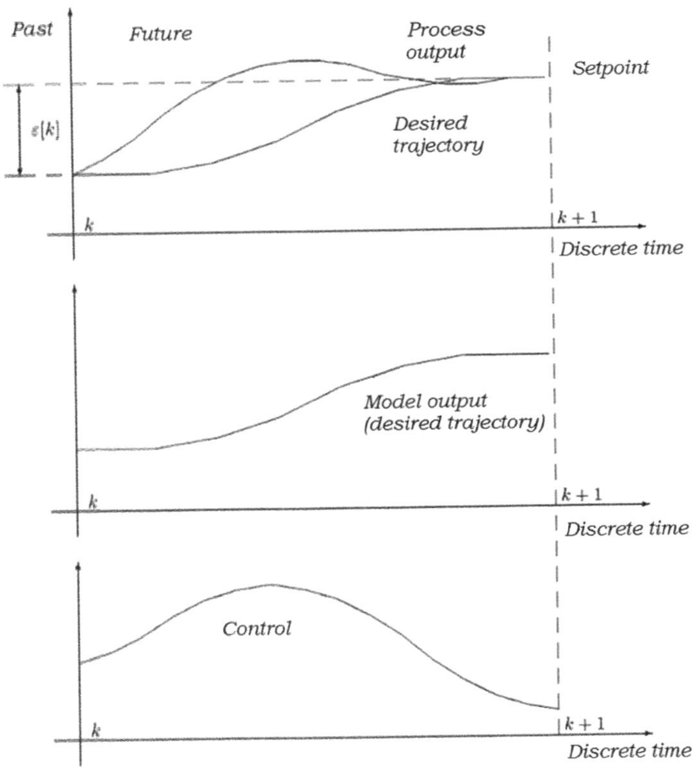

Figure 3.4. *Principle of predictive regulation with a unitary horizon*

The first type of predictive strategy is one that allows the control $u[k]$ to be chosen in such a way that $y[k + 1]$ coincides exactly with the setpoint $r[k]$. We must, therefore, ensure that there is no error at the next sampling instant. This control technique is called dead-beat. It corresponds to the cancelling of the error in one go. Following this condition and equation [3.55], we can write:

$$r[k] = -\sum_{i=1}^{na} a_i y[k-i+1] + \sum_{i=1}^{nb} b_i u[k-i+1], \quad \forall k \in \mathbb{N}, \qquad [3.57]$$

which can be expressed in a compact manner as:

$$\hat{b}(z^{-1})u[z^{-1}] = r[z^{-1}] + \hat{a}(z^{-1})y[z^{-1}]. \qquad [3.58]$$

The predictive control $u[z^{-1}]$ can be calculated following the relation below:

$$u[z^{-1}] = \frac{1}{\hat{b}(z^{-1})} r[z^{-1}] + \frac{\hat{a}(z^{-1})}{\hat{b}(z^{-1})} y[z^{-1}]. \qquad [3.59]$$

Expression [3.60] corresponds to the diagram represented in Figure 3.5.

In the figure, we can identify polynomials R, S, T, with respect to relation [3.60]:

$$T(z^{-1}) = 1, \quad S(z^{-1}) = \hat{b}(z^{-1}), \quad R(z^{-1}) = \hat{a}(z^{-1}). \qquad [3.60]$$

For this type of algorithm, we can see that the following conditions are met: finite stabilization time, minimal rise time and stationary error equal to zero at every sampling instant.

The objective of the control is to adjust the current output $y[k]$ to the value of the reference $r[k]$ in a single step. This objective is not always realistic, as restrictions on the amplitude of the control are not considered. Strong oscillations appear in-between sampling periods, despite an accurate output at those exact periods.

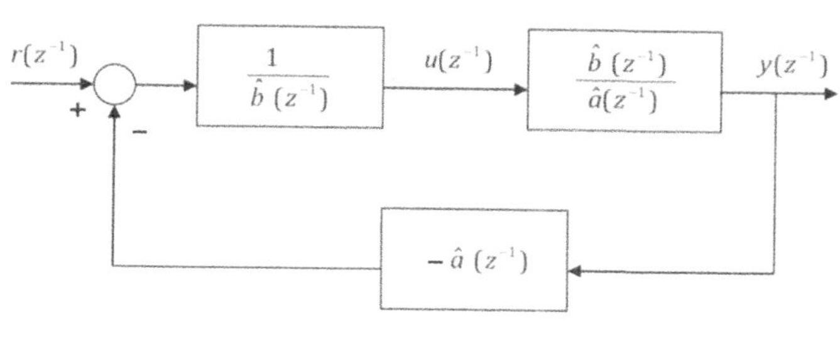

Figure 3.5. *Diagram of predictive dead-beat control*

A more realistic approach involves imposing tracking of the setpoint (seen as the reference trajectory) by the output, following a predetermined transient regime:

$$y[k+1]=y[k]+\alpha\left(r[k]-y[k]\right)=(1-\alpha)\,y[k]+\alpha r[k],\quad \forall k\in\mathbb{N}\,. \qquad [3.61]$$

If $\alpha\in(0,1)$, then the future output belongs to the segment linking $y[k]$ and $r[k]$. For $\alpha=1$, we obtain the previous algorithm. As a result, in [3.61] if we replace the prediction value calculated using [3.56], we get the following recurrent relation:

$$(1-\alpha)y[k]+\alpha r[k]=-\sum_{i=1}^{na}a_i y[k-i+1]+\sum_{i=1}^{nb}b_i u[k-i+1]\,. \qquad [3.62]$$

Using relation [3.61], we can build the current control algorithm:

$$u[z^{-1}]=\frac{\alpha}{\hat{b}(z^{-1})}r[z^{-1}]+\frac{(1-\alpha)+\hat{a}(z^{-1})}{\hat{b}(z^{-1})}y[z^{-1}]\,. \qquad [3.63]$$

We therefore consider that the polynomials of the (RST) canonical structure are defined as follows:

$$\begin{cases} R(z^{-1})=(\alpha-1)-\hat{a}(z^{-1}) \\ S(z^{-1})=\hat{b}(z^{-1}) \\ T(z^{-1})=\alpha \end{cases}. \qquad [3.64]$$

The corresponding implementation diagram is presented in Figure 3.6.

Note the absence of the integral component on the direct path of the predictive structure, which does not guarantee that the stationary error will be equal to zero or that the coefficients of the model will be evaluated accurately.

A predictor expressed in an incremental form eliminates these disadvantages. The incremental predictor is defined from equations [3.55] and [3.56], by subtraction, as follows:

$$\begin{aligned} y[k+1]-y[k] &\overset{def}{=} \sum_{i=1}^{na}a_i y[k-1]-\sum_{i=1}^{na}a_i y[k-i+1] \\ &\quad +\sum_{i=1}^{nb}b_i u[k-i+1]-\sum_{i=1}^{nb}b_i u[k-i] \\ &=(1-z^{-1})\left[\sum_{i=1}^{nb}b_i u[k-i+1]-\sum_{i=1}^{na}a_i y[k-i+1]\right] \\ &=(1-z^{-1})[z^{-d}\hat{b}^{*}(z^{-1})u[k]-\hat{a}^{*}(z^{-1})y[k]],\quad \forall k\in\mathbb{N}\,. \end{aligned} \qquad [3.65]$$

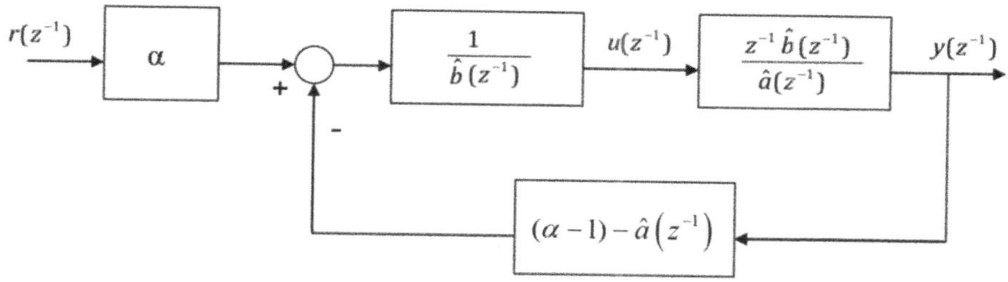

Figure 3.6. *Improved diagram for predictive dead-beat control*

On the other hand, condition [3.63] shows that the practical control algorithm for incremental predictive control is the following:

$$u[z^{-1}]=\frac{\alpha}{(1-z^{-1})\hat{b}(z^{-1})}r[z^{-1}]+\frac{(1-z^{-1})\hat{a}(z^{-1})-\alpha}{(1-z^{-1})\hat{b}(z^{-1})}y[z^{-1}]. \qquad [3.66]$$

This leads to the following polynomials:

$$\begin{cases} R(z^{-1})=\alpha-(1-z^{-1})\hat{a}(z^{-1}) \\ S(z^{-1})=(1-z^{-1})\hat{b}(z^{-1}) \\ T(z^{-1})=\alpha \end{cases}. \qquad [3.67]$$

The diagram corresponding to relation [3.66] is represented in Figure 3.7. In a stationary regime, polynomial $R(z^{-1})$ is reduced to the value α of polynomial $T(z^{-1})$, meaning that the steady-state error has been cancelled out. Moreover, because of the integration effect, this diagram does not require the system parameters to be determined with any great precision as before; in addition, the implementation of this algorithm can be carried out independently of the estimation precision for the parameters of the process model [DAU 04, POP 06].

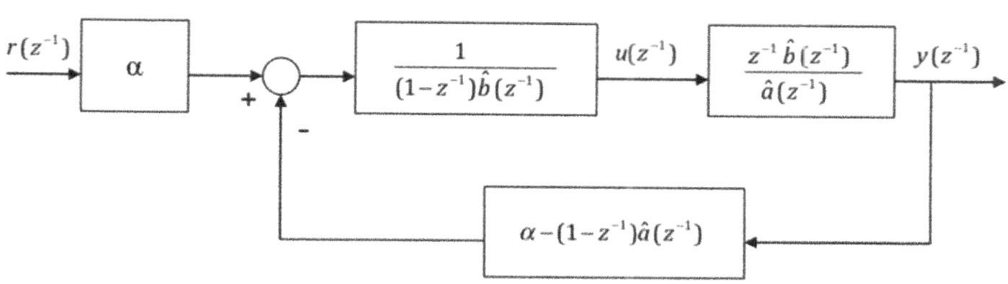

Figure 3.7. *Diagram of incremental predictive control*

4

Adaptive Control and Robust Control

In this chapter, we consider the RST (where RST is R (regulation), S (sensitivity) and T (tracking)) polynomial system structure introduced in section 3.3 for the case where the parameters and the structure of the mathematical model (polynomials $\hat{a}(z^{-1})$ and $\hat{b}(z^{-1})$) are variable over time. In these conditions, the dynamic performance of the closed-loop system needs to be guaranteed using an adaptive or robust strategy [LAN 89, LAN 97, DUG 91, BIT 93, AST 93, AST 95, BES 98, LEF 14].

In the adaptive strategy, the estimation of the model parameters and the design of the control law iteratively succeed one another in real time. Considering the polynomial RST control strategy, the design of its adaptive implementation is based on the recursive estimation of the new parameters in the closed-loop system and on its redesign at every sampling period (of the parameters of the RST polynomials) using recursive adaptation mechanism.

The structure of the nominal system (NS) can vary as well because of nonlinear process behavior or/and uncertainties that occur after the model estimation. In this case, we consider that the discrepancy between the model and the process was due to the time variation both in the structure and the model parameters. Thus, we propose a robust control strategy that ensures tolerance of the system to structural and parametric disturbances of the process. The design of the robust polynomial control improves the nominal control algorithm in order to guarantee the robustness of the system in terms of stability and performance. The RST robust control algorithm will preserve the nominal performances obtained in simulation on the real control system.

Adaptive and robust control systems are extensively used in industrial applications.

4.1. Adaptive polynomial control systems

In order to ensure a unitary point of view and to use the qualities of the RST (where RST is R (regulation), S (sensitivity) and T (tracking)) control, the same representation of the adaptive controller is proposed, meaning the polynomial control structure RST.

Remember that the canonical algorithm is described as follows:

$$u[k]=\frac{T(z^{-1})}{S(z^{-1})}.y^{*}[k]-\frac{R(z^{-1})}{S(z^{-1})}.y[k], \quad \forall k\in\mathbb{N}. \qquad [4.1]$$

The equation that ensures tracking between $r[k]$ and $y[k]$ is written out below:

$$\hat{a}_m(q^{-1})y[k]=\hat{b}_m(q^{-1})r[k]\Leftrightarrow y[k]=y^{*}[k], \quad \forall k\in\mathbb{N}, \qquad [4.2]$$

where $\hat{a}_m(z^{-1})$ and $\hat{b}_m(z^{-1})$ are the polynomials of the trajectory generator evaluated *a priori* (see the structure defined by [3.26]) and $y^*[k]$ is the desired trajectory.

Design of the control regulation (disturbance rejection) for the closed-loop system can be carried out using the method of pole placement presented in section 3.3. Once the parameters of the controller are determined, all there is left to do is to state the method of adaptation to the temporal changes in the model parameters.

4.1.1. *Estimation of the parameters for closed-loop systems*

The adaptation strategy is recommended for the case where the parameters of the model are variable over time (the structure is preserved) and estimation of the parameters of the model is done using an identification technique, for example the method of recursive least squares (RLS), for the closed-loop system. The dynamic model of the system is represented by the equation:

$$\hat{a}_k(z^{-1})y[z^{-1}]=\hat{b}_k(z^{-1})u[z^{-1}], \qquad [4.3]$$

where $\hat{a}_k(z^{-1})$ and $\hat{b}_k(z^{-1})$ are defined by:

$$\begin{cases}\hat{a}_k(z^{-1})\overset{\Delta}{=}1+a_1(k)z^{-1}+\ldots+a_{na}(k)z^{-na}, (\forall)k\in\mathbb{N}\\ \hat{b}_k(z^{-1})\overset{\Delta}{=}b_1(k)z^{-1}+\ldots+b_{nb}(k)z^{-nb}, \quad (\forall)k\in\mathbb{N}\end{cases}. \qquad [4.4]$$

This model can be expressed in the form of a regression:

$$y[k]=\theta^T(k)\varphi[k], \ (\forall)\ k\in\mathbb{N}.$$

Clearly, for any $k \in \mathbb{N}$:

$$\begin{cases} \theta^T(k) = (b_1(k) \quad b_2(k) \ldots b_{nb}(k) \quad a_1(k) \quad a_2(k) \ldots a_{na}(k)) \\ \varphi^T[k] = (u[k-1] \quad u[k-2] \ldots u[k-nb] \quad -y[k-1] \quad -y[k-1] \ldots -y[k-na]) \end{cases}. \quad [4.5]$$

An algorithm of the RLS type can be used to update the parameters of the system using an experimental identification procedure in a closed-loop system (as shown in Chapter 2).

4.1.2. *Design of the adaptive control*

If we choose an adaptation strategy, then the error caused by the variations of the model's parameters, called the *adaptation error*, must be minimized. In practice, this error is defined as follows:

$$e\left[z^{-1}\right] \stackrel{\Delta}{=} \hat{a}_k\left(z^{-1}\right) y\left[z^{-1}\right] - \hat{b}_k\left(z^{-1}\right) u\left[z^{-1}\right], \quad [4.6]$$

often filtered and thus written as:

$$e_f\left[z^{-1}\right] \stackrel{\Delta}{=} \frac{S_k\left(z^{-1}\right)}{P_k\left(z^{-1}\right)} e\left[z^{-1}\right], \quad [4.7]$$

This time the polynomials that defined the RST controller also have coefficients that are variable over time:

$$\begin{cases} R_k(z^{-1}) \stackrel{\Delta}{=} r_0(k) + r_1(k) z^{-1} + \ldots + r_{nr}(k) z^{-nr} \\ S_k(z^{-1}) \stackrel{\Delta}{=} 1 + s_1(k) z^{-1} + \ldots + s_{ns}(k) z^{-ns} \\ T_k(z^{-1}) \stackrel{\Delta}{=} t_0(k) + t_1(k) z^{-1} + \ldots + t_{nt}(k) z^{-nt} \\ P_k(z^{-1}) \stackrel{\Delta}{=} 1 + p_1(k) z^{-1} + \ldots + p_{np}(k) z^{-np} \end{cases} \quad (\forall)\, k \in \mathbb{N}\,. \quad [4.8]$$

Remember that the pole placement method provides a simple relation for the calculation of T, which can also be used for T_k:

$$T_k\left(z^{-1}\right) = \frac{R_k(1)}{P_k(1)} . P_k\left(z^{-1}\right), \quad \forall\, k \in \mathbb{N}. \quad [4.9]$$

Polynomial P_k imposes the poles in a closed-loop system, and polynomials R_k and S_k are calculated as solutions to the following equation:

$$P_k(z^{-1}) = \hat{a}_k(z^{-1})S_k(z^{-1}) + \hat{b}_k(z^{-1})R_k(z^{-1}) . \qquad [4.10]$$

Under these conditions, the adaptive control algorithm can be employed using two different approaches:

– the polynomials R and S are maintained constant and the coefficients of polynomial $T \rightarrow T_k$ (and therefore of polynomial $P \rightarrow P_k$) are made to vary in concordance with the variations of the parameters $\theta(k)$, which constitutes an operation of *tracking adaptation*;

– the characteristic polynomial P and T are maintained constant (and therefore so are the closed-loop poles) and the coefficients of the polynomials $R \rightarrow R_k$, $S \rightarrow S_k$ are made to vary in concordance with the variations of the parameters $\theta(k)$, which constitutes an operation of *adaptation in regulation.*

Following these two approaches, two corresponding adaptive algorithms can be designed. In section 4.2, we shall describe the main steps of these algorithms (after initialization), while pointing out the differences between the two approaches:

1) read the current values of the input $u[k]$, output $y[k]$ and the next value of the setpoint $r[k+1]$;

2) calculate the desired trajectory $y^*[k+1]$ based on the model:

$$G_m\left(z^{-1}\right) = \frac{\hat{b}_m\left(z^{-1}\right)}{\hat{a}_m\left(z^{-1}\right)}; \qquad [4.11]$$

3) evaluate the filtered adaptation error according to a relation that is dependent on the approach used:

i) R, S constant and T, P variable:

$$e_f[z^{-1}] = \frac{S(z^{-1})}{P_k(z^{-1})}(\hat{a}_k(z^{-1})y[z^{-1}] - \hat{b}_k(z^{-1})u[z^{-1}]) ; \qquad [4.12]$$

ii) P constant and R, S, T variable:

$$e_f[z^{-1}] = \frac{S_k(z^{-1})}{P(z^{-1})}(\hat{a}_k(z^{-1})y[z^{-1}] - \hat{b}_k(z^{-1})u[z^{-1}]) ; \qquad [4.13]$$

4) recursively estimate the parameters $\theta(k+1)$ and obtain the new polynomials $\hat{a}_{k+1}(z^{-1})$, $\hat{b}_{k+1}(z^{-1})$ as a result;

5) determine the polynomials R_{k+1}, S_{k+1}, T_{k+1} and P_{k+1} depending on the approach used:

5.1) R, S constant and T, P variable:

$$\begin{cases} R_{k+1}(z^{-1}) = R_k(z^{-1}) \overset{\text{def}}{=} R(z^{-1}) \\ S_{k+1}(z^{-1}) = S_k(z^{-1}) \overset{\text{def}}{=} S(z^{-1}) \\ P_{k+1}(z^{-1}) = \hat{a}_{k+1}(z^{-1})S(z^{-1}) + \hat{b}_{k+1}(z^{-1})R(z^{-1}) \\ T_{k+1}(z^{-1}) = P_{k+1}(z^{-1})\ ; \end{cases} \qquad [4.14]$$

5.2) P, T constant and R, S variable:

$$\begin{cases} P_{k+1}(z^{-1}) = P_k(z^{-1}) \overset{\text{def}}{=} P(z^{-1}) \\ P(z^{-1}) = \hat{a}_{k+1}(z^{-1})S_{k+1}(z^{-1}) + \hat{b}_{k+1}(z^{-1})R_{k+1}(z^{-1}) \Rightarrow R_{k+1}, S_{k+1} \\ T(z^{-1}) = P(z^{-1})\ . \end{cases} \qquad [4.15]$$

6) apply the adaptive control algorithm for the next sampling period $(k + 1)$:

6.1) R, S constant and T, P variable (tracking adaptation):

$$u[z^{-1}] = \frac{1}{S(z^{-1})}(T_{k+1}(z^{-1})y^*[z^{-1}] - R(z^{-1})y[z^{-1}])\ ; \qquad [4.16]$$

6.2) P, T constant and R, S variable (adaptation in regulation):

$$u[z^{-1}] = \frac{1}{S_{k+1}(z^{-1})}(T(z^{-1})y^*[z^{-1}] - R_{k+1}(z^{-1})y[z^{-1}])\ . \qquad [4.17]$$

4.2. Robust polynomial control systems

The robustness of the systems ensures tolerance of the models to structural and parametric disturbances. In industrial applications of control, the robustness of systems is easily measured by the margin of the modulus, defined as the minimal distance between the critical point (meaning the point (–1,0) of the complex plane) and the Nyquist plot of the open-loop system, or by the maximal value of the input–output sensitivity function evaluated over the frequency characteristic of this function [LAN 97, DAU 04, POP 06].

4.2.1. *Robustness of closed-loop systems*

For the nominal system (NS), the Nyquist plot can be traced in the complex plane and the minimal spread of the plot can be evaluated in relation to the critical Nyquist point. We give $|\Delta M(j\omega)|$ as the minimal distance from the Nyquist plot of the open-loop system in relation to the critical Nyquist point (see Figure 4.1). The bigger the margin, the more robust the system is. We can clearly write that:

$$|\Delta M(j\omega)| \overset{\Delta}{=} \min_{\omega \in \mathbb{R}} |1 + G(e^{j\omega})| \qquad [4.18]$$

and $|\Delta M(j\omega)| = \min_{\omega \in \mathbb{R}} |1 + G(e^{j\omega})|$ is the radius of the robustness circle, tangent to the Nyquist plot of the open-loop system.

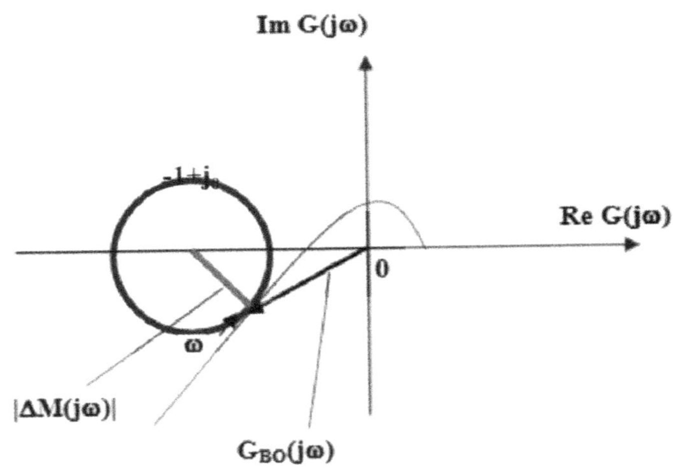

Figure 4.1. *Robustness margin of the system on the Nyquist plot*

We consider the same control structure of the canonical RST type for studying robustness. The polynomial structure of the closed-loop system, in the presence of the random disturbances $v[k]$ and measurement noise $w[k]$, is represented in Figure 4.2.

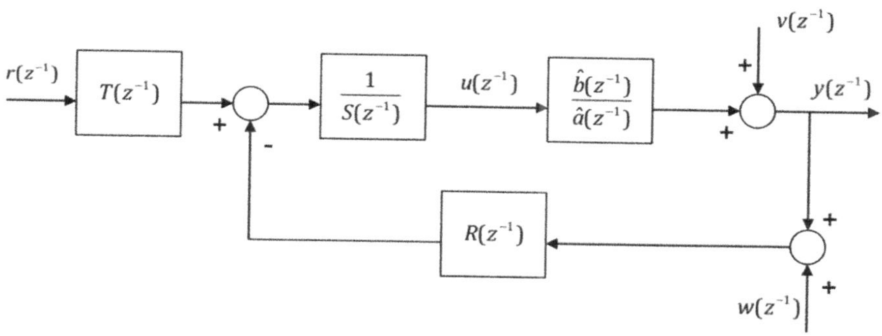

Figure 4.2. *Diagram of RST robust control*

As a result, the output of the system is obtained as follows:

$$y[z^{-1}] = v[z^{-1}] + \frac{\hat{b}(z^{-1})}{\hat{a}(z^{-1})S(z^{-1})}[T(z^{-1})r[z^{-1}] - R(z^{-1})(y[z^{-1}] + w[z^{-1}])]. \quad [4.19]$$

Output y actually verifies the following polynomial equation:

$$\begin{aligned}&\left(\hat{a}(z^{-1})S(z^{-1}) + \hat{b}(z^{-1})R(z^{-1})\right)y[z^{-1}]\\&= \hat{b}(z^{-1})T(z^{-1})r[z^{-1}] + \hat{a}(z^{-1})S(z^{-1})v[z^{-1}] - \hat{b}(z^{-1})R(z^{-1})w[z^{-1}],\end{aligned} \quad [4.20]$$

The influence of the disturbances v plays an important role in studying the robustness of systems. In this context, the analysis of robustness is studied over the transfer function:

$$S_{vy}(z^{-1}) = \frac{\hat{a}(z^{-1})S(z^{-1})}{\hat{a}(z^{-1})S(z^{-1}) + \hat{b}(z^{-1})R(z^{-1})} \quad [4.21]$$

which represents the sensitivity function of the output in relation to the disturbance. In the frequency domain, relation [4.21] becomes:

$$S_{vy}(e^{j\omega}) = \frac{\hat{a}(e^{j\omega})S(e^{j\omega})}{\hat{a}(e^{j\omega})S(e^{j\omega}) + \hat{b}(e^{j\omega})R(e^{j\omega})}, \quad \omega \in \mathbb{R}^+. \quad [4.22]$$

Between the two robustness indicators $|\Delta M(j\omega)|$ and $|S_{vy}(j\omega)|$, there is a clear relationship. Furthermore, it can be noted that:

$$|S_{vy}(e^{j\omega})| = 1/|1+G(e^{j\omega})| \qquad [4.23]$$

and therefore we have

$$|\Delta M(j\omega)| = \min_{\omega\in R}|1+G(e^{j\omega})| = \min_{\omega\in R}\frac{1}{|S_{vy}(e^{j\omega})|} = \frac{1}{\max_{\omega\in R}|S_{vy}(e^{j\omega})|}. \qquad [4.24]$$

In other words, the margin of the modulus $|\Delta M(j\omega)|$ is equal to the inverse of the maximum of the modulus of the sensitivity function $S_{vy}(e^{j\omega})$. In a frequency representation in *dB*, it can be written that:

$$|\Delta M(j\omega)|_{/dB} = -\max_{\omega} S_{vy}(e^{j\omega})_{/dB}. \qquad [4.25]$$

Figure 4.3 shows an usual characteristic of $S_{vy}(e^{j\omega})$ and the procedure for evaluation of $|\Delta M(j\omega)|$. The advantage of designing a robust control is the ability to decrease the maximal value of $S_{vy}(e^{j\omega})$ in order to decrease the effect of random disturbances, and therefore to increase the margin of the robustness modulus. The modulus margin shows both the lower limits of performance in rejecting disturbances as well as the tolerance of the system with regard to its nonlinearity.

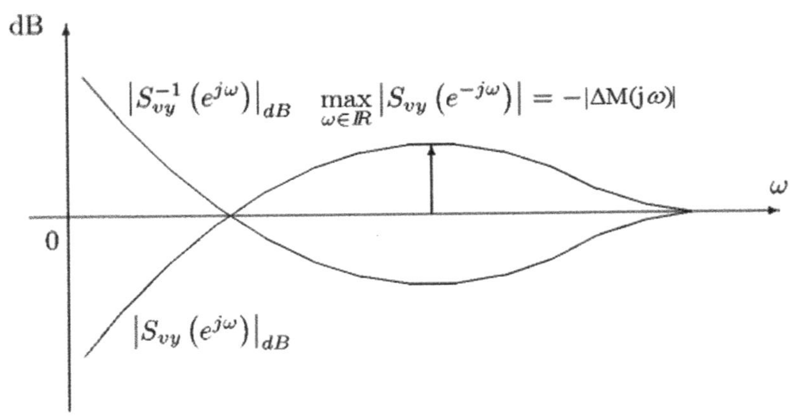

Figure 4.3. *Margin $|\Delta M(j\omega)|$ and the sensitivity function $S_{vy}(e^{j\omega})$*

Before moving onto the design of robust control, the links between the most important properties of the system and its robustness must be studied.

4.2.2. *Studying the stability–robustness connection*

The nominal control ensures the stability and the desired performance level of the NS (*C, M*), which are usually verified in a simulation. Additional conditions must be found that confirm that the real system (*C,P*), driven by a robust control, remains stable, and which therefore guarantee the stability of the system using the robust, closed-loop controller *C*.

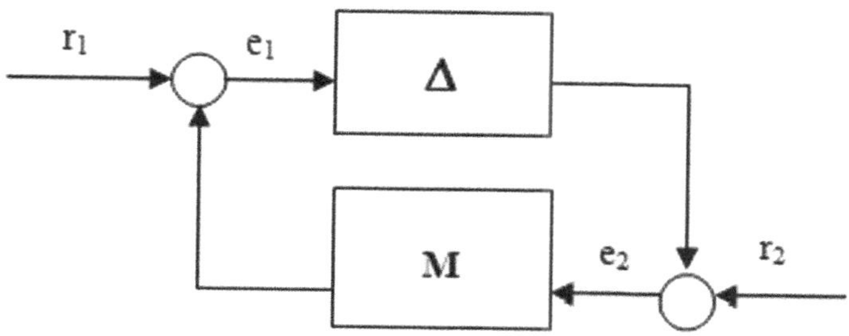

Figure 4.4. *General structure of the system (Δ ,M)*

We consider the generic representation of a closed-loop system *(Δ, M)*, while recalling the conditions of stability imposed by the theorem known as the *small gains theorem*.

Thus, for a stable $M(j\omega)$, the closed-loop structure is stable, for $\Delta(j\omega)$ stable, with $\|\Delta(j\omega)\|_\infty \leq \gamma$, as long as $\|M(j\omega)\|_\infty \leq \gamma^{-1}$, or that

$$\|M(j\omega)\|_\infty . \|\Delta(j\omega)\|_\infty \leq 1, \qquad \omega \geq 0. \qquad [4.26]$$

In the monovariable case, the norm $\|M\|_\infty$ is written as:

$$\|M(j\omega)\|_\infty = \sup_\omega |G(j\omega)|. \qquad [4.27]$$

Let us consider the real closed-loop system (C,P) with multiplicative structural disturbances

$$P = (1+\Delta)\cdot M \, . \qquad [4.28]$$

The structural disturbance is easily deduced verifying the condition $\|\Delta(j\omega)\|_{\infty} \leq \gamma$ of the stability theorem considered. In the multiplicative case, there is:

$$\left|\Delta = \frac{P-M}{M}\right| . \qquad [4.29]$$

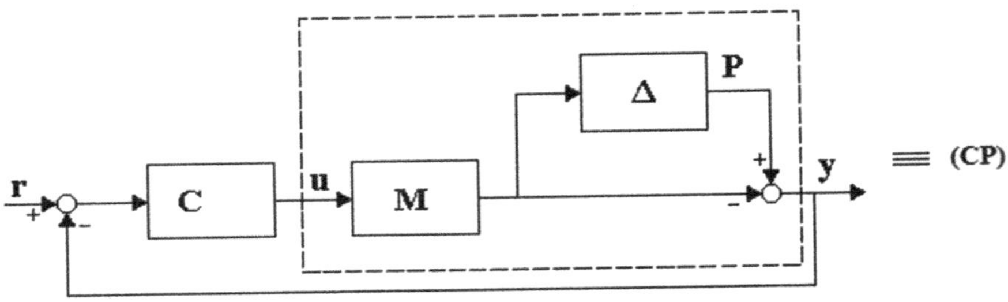

Figure 4.5. *Equivalent structure of the real system (C,P)*

Through some algebraic transformations of the structure presented in Figure 4.5, we obtain the equivalent structure of the real system, in the closed-loop system, as illustrated in Figure 4.6.

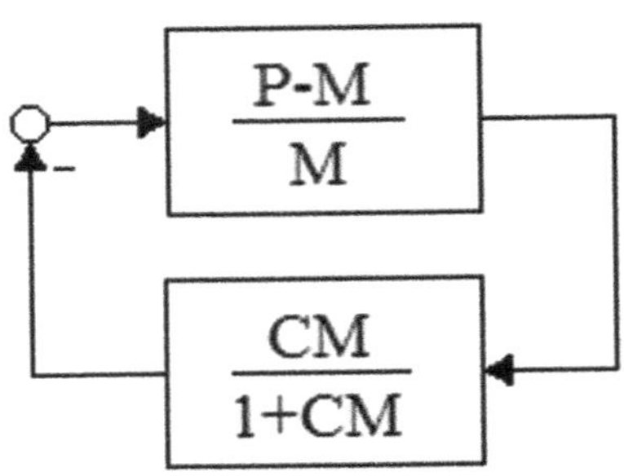

Figure 4.6. *Closed-loop structure of the real system (C,P)*

The "small gains" theorem is applied to the structure of the system [4.6]. The controller C stabilizes the NS *(C, M)*, and the structural disturbance $\Delta(j\omega)$ is bounded:

$$\Delta = \frac{P - M}{M} \leq l_m(j\omega) . \qquad [4.30]$$

If the conditions of the small gains theorem are respected, then controller C stabilizes the real system *(C, P)*, according to the following relation:

$$\left| \frac{P - M}{M} \right| \cdot \left| \frac{C\ M}{1 + C\ M} \right| < 1 . \qquad [4.31]$$

Controller C respects the additional inequality [4.31], and guarantees the stability for the real closed-loop control system *(C, P)*.

4.2.3. *Study of the nonlinearity–robustness connection*

We consider a nonlinear system, with nonlinear input–output characteristics, and whose nonlinearity is located within a conical sector defined by the real and positive values a_1, a_2, where $a_1 > a_2 > 0$ (see Figure 4.7).

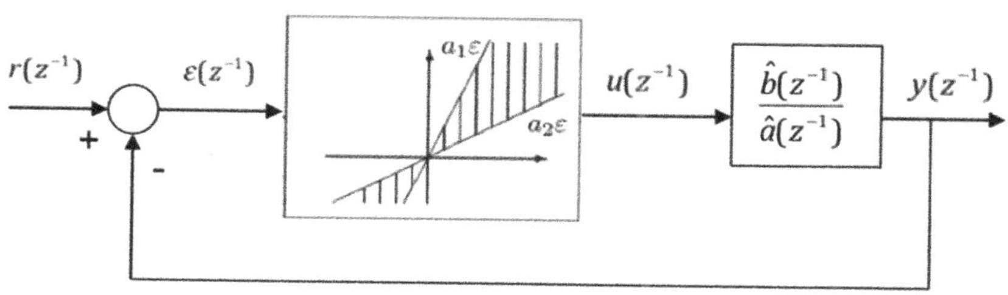

Figure 4.7. *Structure of the nonlinear system*

In order to guarantee the stability of the closed-loop system in the presence of nonlinear components, we must apply the stability criterion (Popov–Zames) for nonlinear systems, the *circle criterion*. The Nyquist plot of $G(e^{j\omega})$ leaves the circle (Popov) to the left, as it is centered on the real axis, which passes through the coordinates $\left(-\frac{1}{a_1}, 0\right)$ and $\left(-\frac{1}{a_2}, 0\right)$ as can be seen in Figure 4.8.

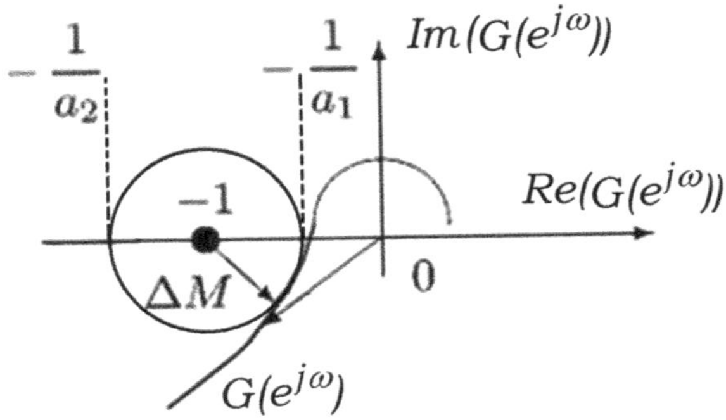

Figure 4.8. *Popov circle and tolerance limits of the nonlinearity*

On the other hand, the modulus margin $|\Delta M(j\omega)|$ defines the robustness circle, of radius $|\Delta M(j\omega)|$, centered in point (–1,0), which also stays left of the Nyquist plot of the open loop transfer function, and tangential to this plot. By superposition of the two circles, it can be noted that the closed-loop system tolerates nonlinearities located inside a conical sector defined by the coordinates $\left(\frac{1}{1-\Delta M},0\right)$ and $\left(\frac{1}{1+\Delta M},0\right)$. This imposes restrictions at the level of the opening of the conical sector, which are expressed by constraints on the values of a_1 and a_2:

$$\frac{1}{1-\Delta M} \geq a_1 \geq a_2 \geq \frac{1}{1+\Delta M}. \qquad [4.32]$$

To conclude, the system whose robustness is measured by the modulus margin $|\Delta M(j\omega)|$ tolerates a nonlinearity located within a conical sector delimited by the values of $\left(\frac{1}{|1+\Delta M(j\omega)|},\frac{1}{|1-\Delta M(j\omega)|}\right)$.

4.2.4. *Study of the performance–robustness connection*

In order to study the performance of the real system and the NS, the closed-loop transfer functions must be considered for the two systems corresponding to (C, P) and (C, M).

The condition of proximity in terms of performance levels between the two dynamic systems is the following:

$$\left|\frac{CP}{1+CP}-\frac{CM}{1+CM}\right|=\text{small}. \qquad [4.33]$$

However, for an easier transformation, condition [4.33] can be factorized and written as

$$\left|\frac{CP}{1+CP}-\frac{CM}{1+CM}\right|=\left|\frac{P-M}{M}\cdot\frac{CM}{1+CM}\cdot\frac{1}{1+CP}\right|=\text{small}. \qquad [4.34]$$

But, the following inequality:

$$\left|\frac{P-M}{M}\right|\cdot\left|\frac{CM}{1+CM}\right|<1 \qquad [4.35]$$

guarantees the robustness of the system in terms of stability; as a result, the final conclusion that ensures robustness in terms of performance levels is the following:

$$\left|\frac{1}{1+CP}\right|=\text{small}. \qquad [4.36]$$

This last result confirms the fact that the disturbance–output sensitivity function (by its maximal value) is an indicator of the robustness of the system; because of this result, which targets the template of the sensitivity function, we next propose to work on the frequency characteristic of this function for the design of robust control.

4.2.5. *Analysis of robustness in the study of the sensitivity function*

The sensitivity function $S_{vy}(e^{j\omega})$ is the central concept for studying the stability and robustness of closed-loop systems. Remember that this indicator is defined using the transfer function of random disturbances v at the output of system y (see [4.22]):

$$S_{vy}(e^{j\omega})\overset{\Delta}{=}\frac{\hat{a}(e^{j\omega})S(e^{j\omega})}{\hat{a}(e^{j\omega})S(e^{j\omega})+\hat{b}(e^{j\omega})R(e^{j\omega})},\quad \omega\in\mathbb{R}^{+}. \qquad [4.37]$$

It is useful at this stage to analyze the most important characteristics of the sensitivity function [4.37].

1) Modulus $|S_{vy}(e^{j\omega})|$ is a measurement of the amplification/attenuation of the disturbance at the output of the system. Thus, we can see that:

– in the frequency zone where $|S_{vy}(e^{j\omega})|=1$ (0 dB), the random disturbances are transferred directly to the output; in this case, the system function as if it were in an "open loop";

– in the frequency zone where $|S_{vy}(e^{j\omega})|<1$ (0 dB), the random disturbances are attenuated, which is the desired scenario;

– in the frequency zone where $|S_{vy}(e^{j\omega})|>1$ (0 dB), the random disturbances are amplified, which is what we are looking to avoid.

2) As the closed-loop system is asymptotically stable, the integral of the logarithm of the modulus $|S_{vy}(e^{j\omega})|$ from 0 to 0.5 f_e (where f_e is the sampling frequency) is equal to 0 for a stable open-loop system:

$$\int_{0}^{0.5 f_e} \log\left|S_{vy}\left(e^{2\pi j f_e}\right)\right| \, df = 0. \qquad [4.38]$$

In other words, the sum of the surfaces between the curve of the sensitivity function and the axis of the frequencies is equal to zero. This implies that the attenuation of the random disturbances in a certain frequency zone leads to their amplification in another zone. This phenomenon is a consequence of the principle of the conservation of energy.

3) The inverse of the maximum of the modulus of the sensitivity function corresponds to the margin of modulus $|\Delta M(j\omega)|$ (a property that has already been demonstrated). The large modulus margin ensures the robustness and the stability of the system, even in the case of systems that present certain nonlinearities.

4) The cancelling out of the effect of random disturbances is obtained (for the polynomial RST control system) in the frequency zone where:

$$\hat{a}\left(e^{j\omega}\right) S\left(e^{j\omega}\right) = 0. \qquad [4.39]$$

Generally speaking, this restriction is assured through the zeros of the polynomial *S*. For example, if *S* contains an integrating element:

$$S\left(e^{j\omega}\right)=\left(1-e^{-j\omega}\right)S'\left(e^{j\omega}\right), \qquad \forall\omega\in\mathbb{R}, \qquad [4.40]$$

then this results in a perfect rejection of the stationary disturbances (for $\omega = 0$).

5) The modulus of the sensitivity function is equal to 1 (0 dB) (for the polynomial RST control system) in the frequency zone where:

$$\hat{b}\left(e^{j\omega}\right)R\left(e^{j\omega}\right)=0\,. \qquad [4.41]$$

As before, this restriction is applied to polynomial *R*. For example, it can be factorized as follows:

$$R\left(e^{j\omega}\right)=\left(1+e^{-j\omega}\right)R'\left(e^{j\omega}\right), \quad \forall\omega\in\mathbb{R}, \qquad [4.42]$$

which ensures a unitary sensitivity function for $\omega = \pi$, which means $f = 0.5f_e$.

6) The introduction of asymptotically stable auxiliary poles $P_a(e^{j\omega})$ generally leads to an attenuation of the modulus of the sensitivity function in the attenuation zones of $1/P_a(e^{j\omega})$.

4.2.6. *Design of the robust RST control*

Robust constraints on stability and performance levels lead to the introduction of a desired *template* for the sensitivity function. With regard to performance levels, the purpose is to ensure rejection of the disturbances in the low-frequency zone and a reduced amplification of the disturbances in the high-frequency zone (the consequence being a small maximal value for modulus $|S_{vy}(e^{j\omega})|$ at high frequencies). The desired performances levels, therefore, require a certain form of the modulus of the sensitivity function. For example, Figure 4.9 illustrates the tube of the template (tolerance template) in which the modulus of the sensitivity function must be found. This tube is determined using a series of specifications imposed *a priori*, and the design of the robust control is then based on the concept of *calibrating* the sensitivity function $S_{vy}(e^{j\omega})$ of the system [LAN 97, DAU 04, POP 06].

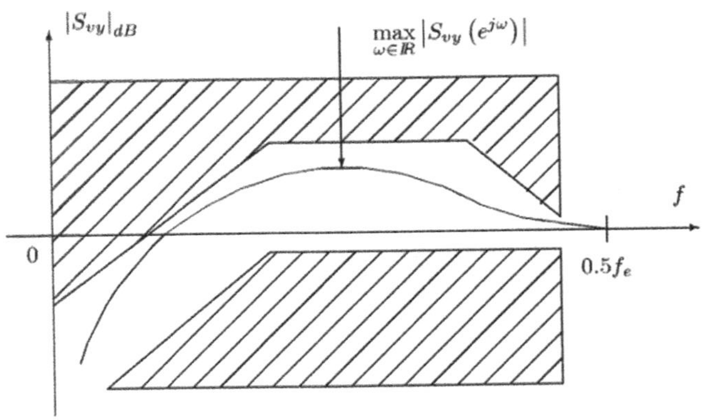

Figure 4.9. *Template tube specified for the sensitivity* $S_{vy}(e^{j\omega})$

The specification of the controller imposes certain zeros and/or poles in the structure of the control in order to decrease sensitivity in relation to the disturbances in a given frequency band. The most common case is to impose the factor $(1 - z^{-1})$ in the structure of polynomial S and/or of factor $(1 + z^{-1})$ in polynomial R. These specifications are generally noted by the auxiliary polynomials:

– $G_R(z^{-1})$: the polynomial specified *a priori* for $R(z^{-1})$:

$$R(z^{-1}) = G_R(z^{-1})R'(z^{-1}) ; \quad [4.43]$$

– $G_S(z^{-1})$: the polynomial specified *a priori* for $S(z^{-1})$:

$$S(z^{-1}) = G_S(z^{-1})S'(z^{-1}) . \quad [4.44]$$

Calculation of the controller is then conducted using the same approach as in the nominal case, after replacing polynomial $\hat{a}(z^{-1})$ with $\hat{a}'(z^{-1}) = \hat{a}(z^{-1})G_S(z^{-1})$ and polynomial $\hat{b}(z^{-1})$ with $\hat{b}'(z^{-1}) = \hat{b}(z^{-1})G_R(z^{-1})$.

4.2.7. *Calibrating the sensitivity function*

In order to decrease the template specified *a priori* around the modulus of the nominal sensitivity function, the dominant poles of the closed loop must be prespecified and filters $G_R(z^{-1})$and $G_S(z^{-1})$ must be also prespecified.

The operation of calibration of the sensitivity function can be formulated in the following manner.

Given an attenuation band, the poles of the closed-loop system must be chosen and polynomials $R(z^{-1})$, $S(z^{-1})$ of the control must be calculated in order to ensure a modulus margin of $|\Delta M(j\omega)|$, while respecting the template imposed on the disturbance to output sensitivity function.

The main idea is therefore to improve the control RST algorithm in order to decrease the maximal value of the sensitivity modulus (to increase the modulus margin of the robustness) using a method for the calibration of the sensitivity function.

The calibration method is an iterative procedure which consists of the following steps:

– choose the dominant poles of P and G_R, G_S imposed by the performance specifications;

– determine the regulator (R' and S');

– test the sensitivity function.

If this remains within the template, then the robust regulator has been accurately determined. Otherwise, there are two possibilities:

– if $\max_{\omega}\left|S_{vy}(e^{j\omega})\right|$ is located within a frequency zone adjacent to the attenuation zone: an extra pair of zeros is introduced in $G_S(z^{-1})$ and/or auxiliary poles $P_a(^{-1})$ in order to decrease the values of $\left|S_{vy}(e^{j\omega})\right|$ in the adjacent zone;

– if $\max_{\omega}\left|S_{vy}(e^{j\omega})\right|$ is located in the high-frequency zone: auxiliary poles are introduced at high frequencies and, if necessary, pairs of zeros can also be introduced (as before in $G_S(z^{-1})$ as well as in $G_R(z^{-1})$).

We then go back to the procedure for the calculation of the control algorithm and this is continued until the template is respected.

In this case, the robust control algorithm is expressed as follows:

$$u\left[z^{-1}\right]=\frac{T\left(z^{-1}\right)}{G_S(z^{-1})S'(z^{-1})}y^*\left[z^{-1}\right]-\frac{G_R(z^{-1})R'(z^{-1})}{G_S(z^{-1})S'(z^{-1})}y\left[z^{-1}\right]. \quad [4.45]$$

5

Multimodel Control

In recent years, many studies on analysis and synthesis problems have been dedicated to the multimodel approach. This was motivated by the desire to consolidate problems in control law synthesis and in the estimation of the state of nonlinear processes with a digital basis. Such opportunities have become possible because of the multimodel approach and through the development of effective digital resolution tools that are mainly based on convex optimization programs.

Automatic control is based on the concept of models that represent the dynamic behavior of a system. For example, a system can be modeled by a mathematical input–output behavior relationship or by an equation that characterizes its change of state. The dilemma then lies between the fidelity of the model to the process and the adequacy of this model in possessing a mathematically and easily exploitable description. Given the difficulty of the task, the engineer is often required to consider some classes of systems, from physical considerations, that follow structural restrictions (linearity, convexity, etc.), which produce approximations of the model. Thus, linear representations have been widely used. The nonlinear model is represented by a linear model obtained as a first approximation within the vicinity of a given point. The disadvantage of this approach is its uniquely local validity.

To overcome this difficulty, a comprehensive approach was proposed: this was based on multiple models that were defined around different operating points. This approach, called multimodel, is a representation that can be obtained either from an initial nonlinear model or directly from input–output data.

Command models are evaluated multiple times throughout a linearization around an operating point. Moreover, the process has a nonlinear characteristic and this model is difficult to exploit. Therefore, one should use representations that are each valid in a limited field of evolution. To achieve the same level of performance at different operating points for the nonlinear feature, the solution put forward in this chapter involves using the notion of multimodel command based on a linguistic approach.

5.1. Construction of multimodels

The design of a process control is based on a model that is usually obtained through observation of its behavior [CHA 12].

Most often, this is a mathematical relationship between the inputs and outputs of the process.

As this model is difficult to use due to its complexity or its validity that is limited to a given domain of evolution, it is then appropriate to use representations that are each valid within a limited domain of evolution.

An approach that performs well and is easy to implement involves a linguistic representation using fuzzy logic [ZAD 65] and Mamdani's approach [MAM 75].

The Takagi–Sugeno representation [TAK 85], which is also based on fuzzy logic, implements a set of process models that are usually defined around various operating points.

The most commonly used representation is one based on a set of linearized models. These models can be obtained either from an initial nonlinear model or directly from input–output observations such as in the neuronal approach.

5.1.1. *Fuzzy logic: Mamdani models*

The information available on the process includes a certain number of variables that either correspond to measurements determined for the system or set points that define the desired behavior of the process and which are determined by experience.

Each of these variables is assigned various possible membership classes such as "small, medium or large". In general, the number of classes is limited to five and the field of possible evolution of each variable, called the universe of discourse, must be completely encompassed by all classes and a membership coefficient $\mu_c(v) \in [0 \quad 1]$ of variable v for class C is attributed.

$\mu_c(v) = 1$ means that variable v totally belongs to class C and $\mu_c(v) = 0$ means that variable v does not at all belong to class C.

The implementation of the fuzzy approach generally requires partial recovery of two classes that correspond to values in the vicinity of variable v in addition to the total recovery of the universe of discourse by all classes, meaning the set of values admissible by variable v (Figure 5.1).

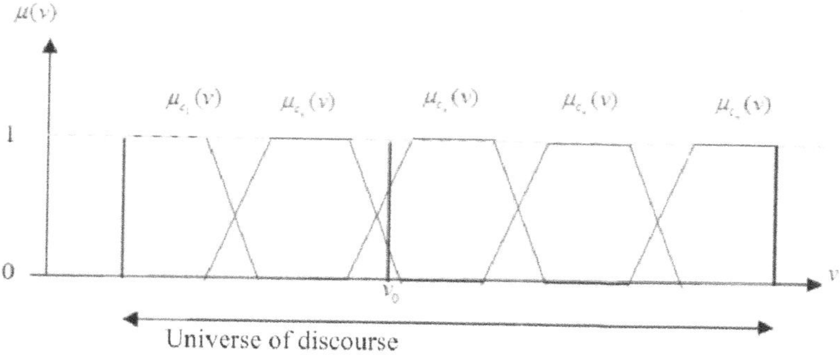

Figure 5.1. *Division into classes*

For example, for value v_0 indicated in Figure 5.1, variable v partly belongs to both classes C_2 and C_3.

This division into classes of permissible values for variable v is called fuzzification.

Similarly, the set of possible values for the decision variables, so those that allow process evolution, is divided into classes.

The definition of the model results from the expertise of the user who associates a decision class from a set of rules called an "inference engine" for each situation class that is defined by the set of values from the data available.

Each data set that defines several condition classes is associated with one or more decision classes. The exact value of each decision variable is then obtained through a process called defuzzification (see Figure 5.2).

Let us suppose that the data correspond to the variable $\mathbf{X}_0 = \begin{bmatrix} \mathbf{x}_{0_1}, & \mathbf{x}_{0_2} & \cdots & \mathbf{x}_{0_n} \end{bmatrix}^T$ and the decisions to be made correspond to the variable $\mathbf{Y}_0 = \begin{bmatrix} \mathbf{y}_{0_1}, & \mathbf{y}_{0_2} & \cdots & \mathbf{y}_{0_m} \end{bmatrix}^T$. If $\mathbf{x}_{0_i}$ partially belongs to class $\mathbf{X}^i_{k_i}$ with the membership coefficient $\mu^i_{k_i}(\mathbf{x}_{0_2})$, we would say that $\mathbf{x}_{0_i}$ is $\mathbf{X}^i_k$. Based on some base rules (provided by an expert) and the fuzzy subsets that correspond to the fuzzification of the vector of data $\mathbf{X}_0$, the inference mechanism determines the fuzzy subsets that comprise the decision variable $\mathbf{Y}_0$.

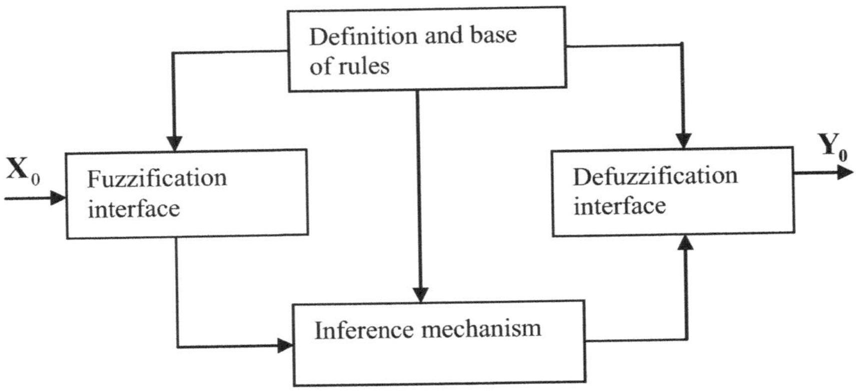

Figure 5.2. *Structure of fuzzy control*

The rule base is:

If $\mathbf{x}_{0_1}$ is $\mathbf{X}^1_{k_1}$, $\mathbf{x}_{0_2}$ is $\mathbf{X}^2_{k_2}$,... $\mathbf{x}_{0_{kn}}$ is $\mathbf{X}^n_{k_n}$, then $\mathbf{y}_{0_j}$ is $\mathbf{Y}^j_{l_p}$.

The rules of the base are applied in a way that defines the fuzzy sets that each of the decision variables $\mathbf{y}_{0_j}$ belong to.

If $\mu^i_{k_l}(\mathbf{x}_{0_i})$ is the membership coefficient of $\mathbf{x}_{0_1}$ to class $\mathbf{X}^i_{k_1}$, it follows that for the membership coefficient $\mathbf{y}_{0_j}$ to class $\mathbf{Y}^j_{l_p}$, we get:

$$\mu^j_{l_p}(\mathbf{y}_{0_j}) = \mu^1_{k_1}(\mathbf{x}_{0_1}) \wedge \mu^2_{k_2}(\mathbf{x}_{0_2}),\ldots,\wedge\mu^n_{k_n}(\mathbf{x}_{0_n}), \qquad [5.1]$$

and for the application of all rules r_j from the rule base used, for the *j*th decision variable, we get:

$$\mu_{rj}(\mathbf{y}_{0j}) = \mu^j_{l1}(\mathbf{y}_{0j}) \vee \mu^j_{l2}(\mathbf{y}_{0j}),\ldots, \vee\, \mu^j_{lrj}(\mathbf{y}_{0j}). \qquad [5.2]$$

In writing, $\wedge$ represents the logical operator AND and $\vee$ represents the logical operator OR.

The most used logical operators in the context of multimodel representations are “min” for AND and “max” for OR.

The most commonly used defuzzification operator is the barycentric operator, so for variable z of membership coefficient $\mu(\mathbf{z})$:

$$D_F(\mathbf{z}) = \mathbf{z}_0 = \frac{\int_{D\mathbf{z}} \mathbf{z}\mu(\mathbf{z})d\mathbf{z}}{\int_{D\mathbf{z}} \mu(\mathbf{z})d\mathbf{z}}. \tag{5.3}$$

In this way, $D_\mathbf{z}$ represents the field of evolution of variable **z**.

An example of application of the production of a fuzzy PI control is given in [BOR 98].

Let us consider output process **y** and control u, for which only the domains of evolution of various measured variables of the command to be carried out are known.

The instant input–output gain of the process is positive: **y** increases if u increases and **y** decreases if u decreases.

Let us note the reference variable $\mathbf{y}_c$,

$$\varepsilon = \mathbf{y}_c - \mathbf{y}, \dot{\varepsilon} = \frac{d\varepsilon}{dt} = -\frac{d\mathbf{y}}{dt} \text{ (if } \mathbf{y}_c = \text{constant)}. \tag{5.4}$$

Three classes are defined for ε and $\dot{\varepsilon}$, five classes for Δu, variation of u, as shown in Figure 5.3.

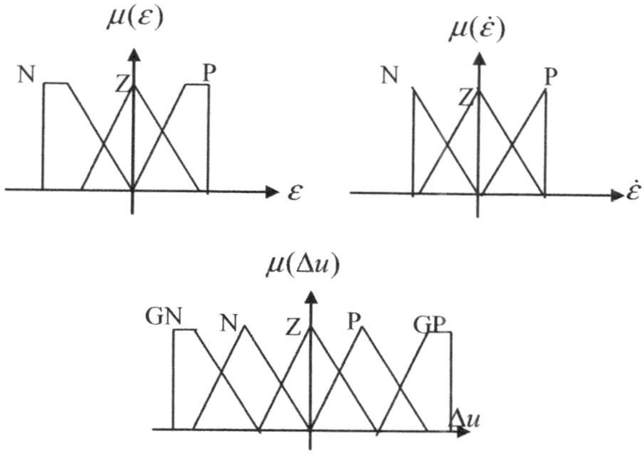

Figure 5.3. *Data classes*

Given the properties of the process, the table of rules can be built by simply taking the instantaneous data into account.

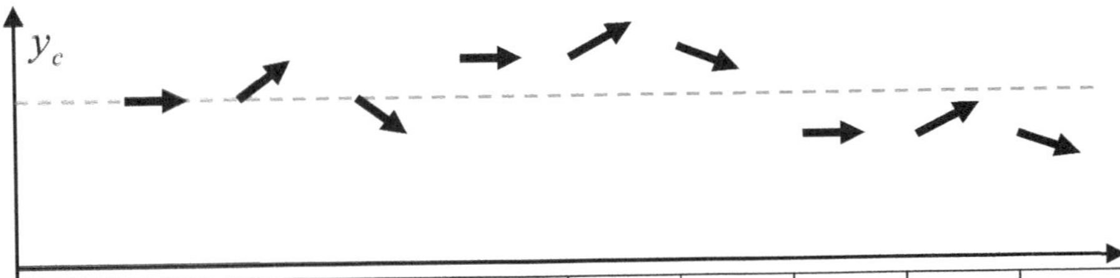

ε	Z	Z	Z	N	N	N	P	P	P
$\dot{\varepsilon}$	Z	N	P	Z	N	P	Z	N	P
Δu	Z	N	P	N	GN	Z	P	Z	GP

Figure 5.4. *Establishment of rules*

$\dot{\varepsilon}$ \ ε	N	Z	P
N	GN	N	Z
Z	N	Z	P
P	Z	P	GP

Figure 5.5. *Table of rules defining $\Delta\mu$*

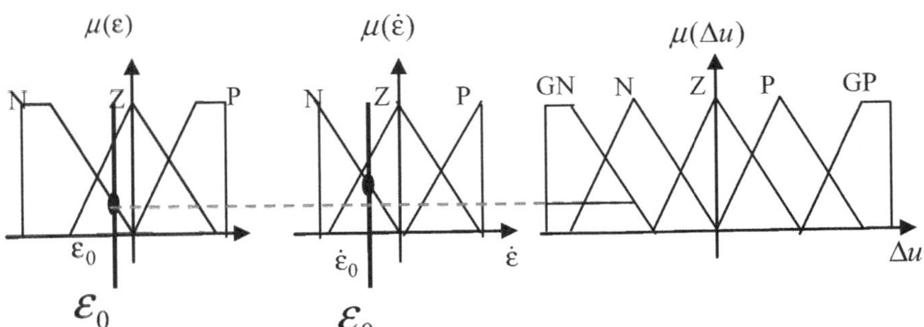

ε is N and $\dot{\varepsilon}$ and $\dot{\varepsilon}$ is N $\Rightarrow$ Δu is GN

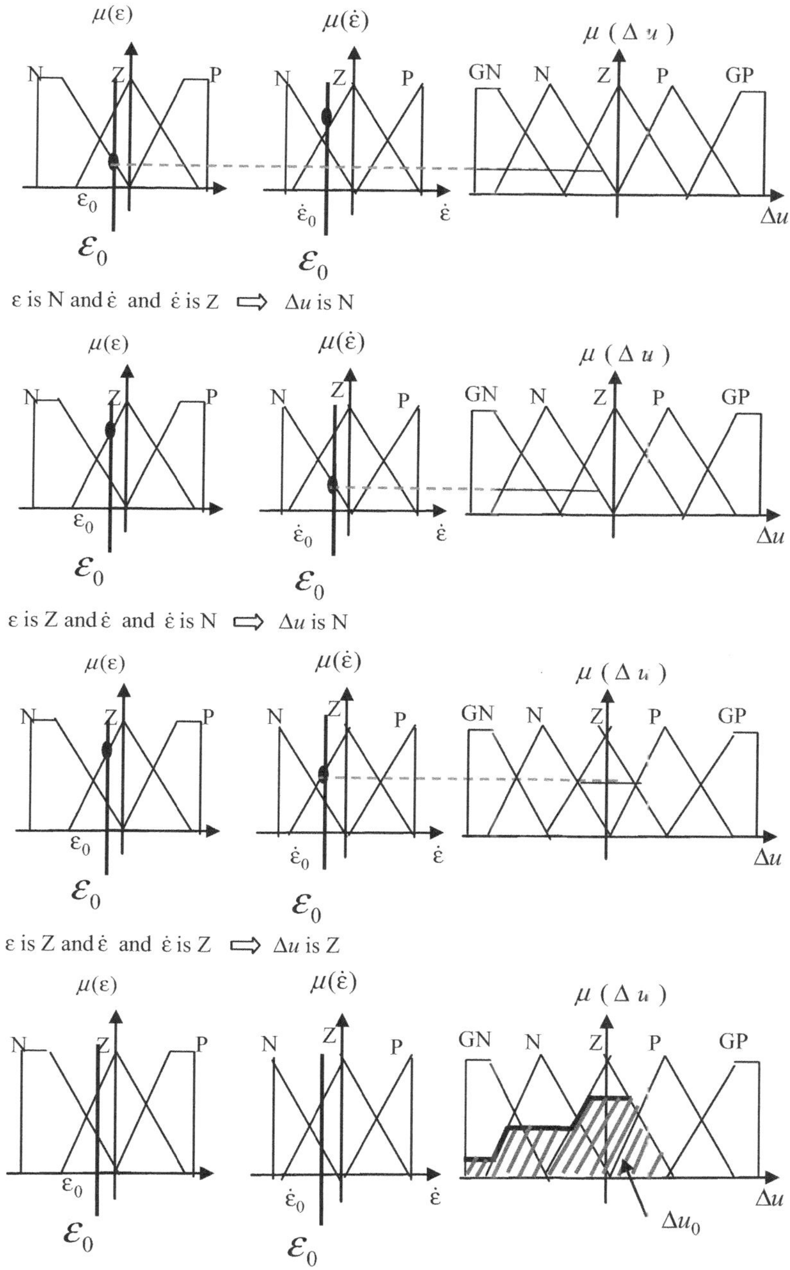

Figure 5.6. *Defuzzification*

The value of Δu_0 of command variation is the abscissa of the center of gravity of the hatched area in the figure obtained through the OR (the maximum) of various situations.

5.1.2. *Identification from input–output data: direct method*

Linear models are identified from input–output data in the vicinities of various operating points.

The most commonly used linear model identification methods with unknown parameters are based on the minimization of a functional deviation between the estimated output $\mathbf{y}_m(t)$ of the model and the measured output $\mathbf{y}(t)$ of the process.

$$J(\theta)=\frac{1}{2}\sum_{k=1}^{h}\varepsilon^2(k,\theta)=\frac{1}{2}\sum_{k=1}^{h}(y_m(k)-y(k))^2 \ . \qquad [5.5]$$

In this expression, h is the horizon of observation and θ is the vector of the model parameters.

The parameter update algorithm is:

$$\theta(k+1)=\theta(k)-\eta D(k)\ , \qquad [5.6]$$

where k is the iteration index, $\theta(k)$ is the value of the parameter vector for the kth iteration and η is an adjustment factor for modifying the speed of convergence. $D(K)$ is the search direction depending on the optimization method used to minimize $J(\theta)$.

The most frequently used optimization methods are:

– the gradient algorithm

$$D(k)=G(k)=\left.\frac{\partial j}{\partial\theta}\right|_{\theta=\theta(k)}=\sum_{k=1}^{h}\frac{\partial\varepsilon(k,\theta)}{\partial\theta}\varepsilon(k,\theta)\Big|_{\theta=\theta(k)}\ ; \qquad [5.7]$$

– the Newton algorithm

$$D(k)=\mathbf{H}^{-1}(k)G(\theta k))\ ,$$

where **H**(*k*) is the Hessian matrix of the criterion

$$\mathbf{H}(k)=\sum_{k=1}^{h}\frac{\partial\varepsilon(k,\theta)}{\partial\theta}\frac{\partial\varepsilon(k,\theta)}{\partial\theta^{T}}+\sum_{k=1}^{h}\frac{\partial^{2}\varepsilon(k,\theta)}{\partial\theta^{2}}\bigg|_{\theta=\theta(k)}; \qquad [5.8]$$

– the Gauss–Newton algorithm, which is a simplified form of the Newton algorithm:

$$\mathbf{H}_{a}(k)=\sum_{k=1}^{h}\frac{\partial\varepsilon(k,\theta)}{\partial\theta}\frac{\partial\varepsilon(k,\theta)}{\partial\theta^{T}}_{\theta=\theta(k)}. \qquad [5.9]$$

5.1.3. *Identification from input–output data: neural approach*

In theory, this method optimizes the number of models to be used in the multimodel method and determines the set of data to be considered for the determination of each model.

Once all the data are collected, they are grouped into classes by an artificial neural network for competitive learning with the rival penalty [XU 93, BOR 07, ELF 08]. The neural network used comprises a single hidden layer.

Let us note that

$$\mathbf{d}(t)=\left[\mathbf{d}_{1}(t),\quad \mathbf{d}_{2}(t),\quad \ldots\quad \mathbf{d}_{v}(t)\right]^{T} \qquad [5.10]$$

is the vector containing all the data available at time *t*

$$\mathbf{z}^{(1)}(t)=\left\|\mathbf{w}_{j}(t)-\mathbf{d}(t)\right\|, \qquad [5.11]$$

where $\mathbf{w}_{j}(t)=\left[\mathbf{w}_{j1}(t),\quad \mathbf{w}_{j2}(t),\quad \ldots\quad \mathbf{w}_{jv}(t)\right]^{T}$ is the vector of the weights connecting inputs at the *j*th neuron. This vector corresponds to class C_j

– $\mathbf{z}_{j}^{(2)}(t)$ output of the *j*th neuron in the output layer;

– $\mathbf{z}^{(2)}(t)=\arg\min_{j}\left\|\mathbf{w}_{j}(t)-\mathbf{d}(t)\right\|$.

Starting from *K* number of arbitrarily selected classes where they are each represented by a vector $\mathbf{w}_j$, the data vectors are successively presented to the network.

Let us note that:

– $g = \arg\min_{k} \left\| \mathbf{w}_k(t) - \mathbf{d}(t) \right\|$;

– $r = \arg\min_{k \neq g} \left\| \mathbf{w}_k(t) - \mathbf{d}(t) \right\|$.

The updating of weights at each presentation of data $\mathbf{d}(t)$ is as follows:

$$\begin{aligned} \mathbf{w}_g(t+1) &= \mathbf{w}_g(t) + \eta(t)(\mathbf{d}(t) - \mathbf{w}_g(t)) \\ \mathbf{w}_r(t+1) &= \mathbf{w}_r(t) - \gamma(t)(\mathbf{d}(t) - \mathbf{w}_r(t)) \end{aligned} \qquad [5.12]$$

with the other weights remaining unchanged.

The learning coefficients $\eta(t)$ and $\gamma(t)$ belong to the interval $]0,1[$ and $\gamma(t)$ is much lower than $\eta(t)$.

After repeatedly, if necessary, presenting the various data to the network, the latter can determine the number of classes present, indeed:

– if the K number of output neurons is chosen to be higher than the number of classes present, the winning neurons move toward the class centers and the neurons in excess move away from the set of observations;

– if the K number of output neurons proves to be less than the number of classes, there will be an oscillation of the weight vectors between the various classes, which requires adding one or more classes.

Once the optimum number K of classes has been determined, refinement of the grouping of the various data classes may be carried out by following the Kohonen rule.

Based on the class centers that were previously obtained, the classification is performed by the algorithm:

$$\begin{aligned} g &= \arg\min_{k} \left\| \mathbf{w}_k(t) - \mathbf{d}(t) \right\| \\ \mathbf{w}_g(t+1) &= \mathbf{w}_g(t) + \eta(t)(\mathbf{d}(t) - \mathbf{w}_g(t)) \end{aligned} \qquad [5.13]$$

by successively presenting the various data to the network, the other weights remain unchanged.

Each of the K basic models is then identified from each of the previously identified data sets.

5.1.4. *Linearization around various operating points*

If one has a nonlinear mathematical model of the process, the model base can be achieved by linearization of this model around various operating points.

Thus, the model:

$$\dot{\mathbf{x}}(t)=\mathbf{f}(\mathbf{x}(t),u(t)), \qquad [5.14]$$

where $\mathrm{f}:\mathbb{R}^n\times\mathbb{R}^m\rightarrow\mathbb{R}^n$ of class C^1.

In the vicinity of the point $(\mathbf{x}_i,u_i)$, by the first-order development, we can write

$$\dot{\mathbf{x}}(t)=\mathbf{A}_i\mathbf{x}(t)+\mathbf{B}_iu(t)+\mathbf{d}_i, \qquad [5.15]$$

where

$$\mathbf{A}_i=\left.\frac{\partial\mathbf{f}(\mathbf{x},u)}{\partial\mathbf{x}}\right|_{\substack{x=x_i\\u=u_i}},\mathbf{B}_i=\left.\frac{\partial\mathbf{f}(\mathbf{x},u)}{\partial u}\right|_{\substack{x=x_i\\u=u_i}} \qquad [5.16]$$

$$\mathbf{d}_i=\mathbf{f}(\mathbf{x}_i,u_i)-\mathbf{A}_i\mathbf{x}_i-\mathbf{B}_iu_i. \qquad [5.17]$$

In the case of *k* local models, we obtain the multimodel formula:

$$\dot{\mathbf{x}}(t)=\sum_{i=1}^{k}\eta_i(\mathbf{z}(t))(\mathbf{A}_i\mathbf{x}(t)+\mathbf{B}_iu(t)+\mathbf{d}_i), \qquad [5.18]$$

where $\eta_i(\mathbf{z}(t))$ are the activation functions and $\mathbf{z}(t)$ is the vector of measurable variables of the system.

5.1.5. *Convex polytopic transformation from an analytical model refined for the command*

The nonlinear model evolution is governed by the equations [CHA 12]:

$$\begin{aligned}\dot{\mathbf{x}}(t)&=\mathbf{f}(\mathbf{x}(t))+\mathbf{B}(\mathbf{x}(t))u(t)\\ \mathbf{y}&=\mathbf{g}(x(t))\end{aligned}, \qquad [5.19]$$

where $\mathbf{x} \in \mathbb{R}^n, u \in \mathbb{R}^m, \mathbf{y} \in \mathbb{R}^l, \mathbf{f} \in \mathbb{R}^n, \mathbf{g} \in \mathbb{R}^l, \mathbf{B} \in \mathbb{R}^{n \times m}$. If the various nonlinearities involved in the model are continuous and bounded, a convex polytopic transformation of each nonlinear element enables the systematic definition of the model base [CHA 02, CHA 02, MOR 00].

Indeed, let $\mathbf{h}(\mathbf{v}(t))$ be a bounded function for all $\mathbf{v} \in [a,b]$

Then, there are two functions:

$$\mathbf{h}^i(\mathbf{v}(t)) : [a,b] \rightarrow [0,1], i = 1,2,$$

where $\mathbf{h}^1(\mathbf{v}(t)) + \mathbf{h}^2(\mathbf{v}(t)) = 1$ and two scalars α and β such that

$$\mathbf{h}(\mathbf{v}(t)) = \alpha \mathbf{h}^1(\mathbf{v}(t)) + \beta \mathbf{h}^2(\mathbf{v}(t)). \qquad [5.20]$$

For example, by taking

$$\alpha = \max_{v \in [a,b]} \left(\mathbf{h}(\mathbf{v}(t))\right), \beta = \min_{v \in [a,b]} \left(\mathbf{h}(\mathbf{v}(t))\right). \qquad [5.21]$$

We can choose:

$$\mathbf{h}^1(\mathbf{v}(t)) = \frac{\mathbf{h}(\mathbf{v}(t)) - \beta}{\alpha - \beta}, \mathbf{h}^2(\mathbf{v}(t)) = \frac{\alpha - \mathbf{h}(\mathbf{v}(t))}{\alpha - \beta}. \qquad [5.22]$$

The model is written as:

$$\begin{pmatrix} \dot{\mathbf{x}}(t) \\ \mathbf{y}(t) \end{pmatrix} = \begin{bmatrix} \mathbf{A}(\mathbf{x}(t)) & \mathbf{B}(\mathbf{x}(t)) \\ \mathbf{C}(\mathbf{x}(t)) & 0 \end{bmatrix} \begin{bmatrix} \mathbf{x}(t) \\ u(t) \end{bmatrix} \qquad [5.23]$$

letting

$$\mathbf{E}(\mathbf{x}(t)) = \begin{bmatrix} \mathbf{A}(\mathbf{x}(t)) & \mathbf{B}(\mathbf{x}(t)) \\ \mathbf{C}(\mathbf{x}(t)) & 0 \end{bmatrix} \qquad [5.24]$$

if $\mathbf{E}(\mathbf{x}(t))$ is continuous and bounded, by considering each non-constant term, the matrix $\mathbf{E}(\mathbf{x}(t))$ can be transformed by:

$$\mathbf{E}(\mathbf{x}(t)) = \sum_{i=1}^{N} \eta_i(\mathbf{x}(t)) \mathbf{E}_i, \quad \mathbf{E}_i = \begin{bmatrix} \mathbf{A}_i & \mathbf{B}_i \\ \mathrm{C}_i & 0 \end{bmatrix}. \qquad [5.25]$$

Number N of the model base from the transformation depends on number s of nonlinearities contained in the matrix $\mathbf{E}(\mathbf{x}(t)) : N = 2^s$.

We see that this approach does not introduce any approximation error [CHA 12].

5.1.6. *Calculation of the validity of base models*

Between 0 and 1, validity characterizes the relevance at every moment of every model [DEL 96].

Let us note that $\mathbf{z}$ is the vector of the measurable variables of the process and $\mathbf{z}_i$ is the vector corresponding to the ith model. We let:

$$\mathbf{r}_i = \|\mathbf{z} - \mathbf{z}_i\|, i = 1, \ldots, k \qquad [5.26]$$

The chosen norm is the Euclidean norm.

$$\mathbf{r}_i^{'} = \frac{\mathbf{r}_i}{\sum_i^k \mathbf{r}_i}, \mathbf{v}_i^{'} = 1 - \mathbf{r}_i^{'}. \qquad [5.27]$$

We get an initial expression of validity (simple validity):

$$\mu_i^{simp} = \frac{\mathbf{v}_i^{'}}{k-1}. \qquad [5.28]$$

If the data classes for the various models are well separated, reinforced validities can be defined as [ELF 08]:

$$\begin{aligned} \mathbf{v}_i^{''} &= \mu_i^{simp} \prod_{i=1}^{k} (1 - \mu_i^{simp}) \\ \mu_i^{simp} &= \frac{\mathbf{v}_i^{''}}{\sum_{i=1}^{k} \mathbf{v}_i^{''}} \end{aligned} \qquad [5.29]$$

In the implementation of multimodels, we sometimes use the terms “membership function” or “activation function” instead of the term “validity”.

5.2. Stabilization and control of multimodels

Multimodel control is achieved by separately calculating the command for each of the models or submodels.

Let us consider the model M_i of command $u_i(.)$ and activation coefficient $\eta_i(.)$

with $\eta_i(.) \geq 0, \quad \sum_i \eta_i(.) = 1$.

For the multimodel, we get command:

$$u(.) = \sum_{i=1}^{k} \eta_i(.) u_i(.) . \qquad [5.30]$$

This approach is particularly effective and remains valid even if the base models and their commands are of different natures and orders.

For example, far from the target, one could consider a command with an optimality criterion of the type minimum time or minimum energy and for the model closer to the target, a proportional and integral command could be considered.

When the base models are of the same type and same order, we denote θ as the vector of parameters that determine the command system.

Under these circumstances, if θ_i is the vector of correction parameters associated with model M_i, we can take the multimodel correction parameters as the vector:

$$\theta(.) = \sum_{i=1}^{k} \eta_i(.) \theta_i \qquad [5.31]$$

and this goes regardless of the nature of the correction used.

5.3. Design of multimodel command: fuzzy approach

When all models are of the same order and defined in the state space with a similar state vector, the Takagi–Sugeno approach [TAK 85] is often used.

In this case, the main command systems correspond to a state feedback or an output feedback such as a command by poles placement, an RST command (mainly for sampled processes) and a command with quadratic criterion.

Let us consider the state equation model M_i :

$$\begin{aligned}\dot{\mathbf{x}}(t) &= \mathbf{A}_i(\mathbf{x}(t)+\mathbf{B}_i u_i(t)\\ \mathbf{y}(t) &= \mathbf{C}_i(\mathbf{x}(t))\end{aligned} \quad [5.32]$$

Command u_i is then written as:

$$u_i(t) = K_i\mathbf{x}(t)) + l_i\mathbf{y}^c(t), \quad [5.33]$$

where $\mathbf{y}^c(t)$ designates the desired output.

The multimodel command is then:

$$u(t) = \sum_{i=1}^{k} \eta_i(t) u_i(t) \quad [5.34]$$

or

$$u(t) = \left(\sum_{i=1}^{k} \eta_i(.)(\mathbf{A}_i + \mathbf{B}_i K_i)\right)\mathbf{x}(t) + \left(\sum_{i=1}^{k} \eta_i(.)\mathbf{B}_i l_i\right)\mathbf{y}(t). \quad [5.35]$$

All multimodel command techniques are discussed in detail in [CHA 12].

5.4. Trajectory tracking

Multimodel command is particularly suited to the trajectory tracking problem in a nonlinear system. Indeed, we just need to choose a set of points on the desired trajectory and determine a linearized model for each of these points.

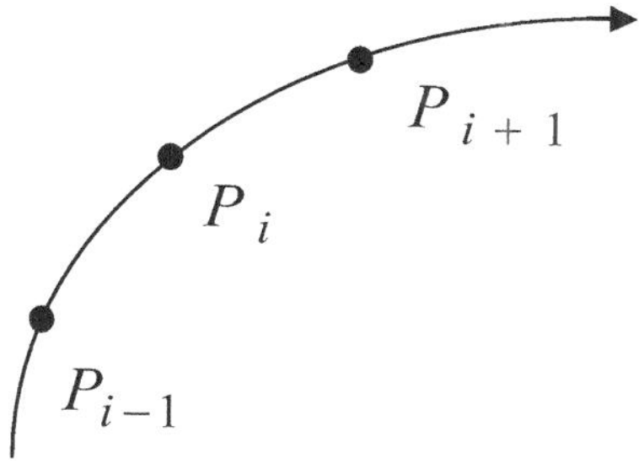

For example, starting at point P_i, the command u_{i+1} to get to point P_{i+1} is determined from a model M_{i+1} that is defined at this point or to avoid encountering problems linked to switching through fusion of commands, u_i and u_{i+1} are calculated to get from point P_i to point P_{i+1}, with u_i being calculated from the model M_i defined at point P_i and u_{i+1} being calculated from the model M_{i+1} defined at point P_{i+1}:

$$u_{i,i+1} = \eta_i u_i +_i \eta_{i+1} u_{i+1} \,. \qquad [5.36]$$

Given the nonlinearities and uncertainties in the process model, it is possible that point P_{i+1} is not reached but rather that the process evolution made it go within the vicinity of this point, an area that can be identified by calculating an attractor, as defined in Chapter 6.

6

Ill-Defined and/or Uncertain Systems

To calculate command laws for a process, we use a model. If the model used is ill-defined and/or uncertain with bounded uncertainties or disturbances, we then need to check if the process evolves in accordance with the prescribed law with an error that is consistent with the desired behavior.

A particularly effective approach to this problem is to study the stability of the command process and estimate the maximum error in process behavior relative to that desired. This can be done using the concept of a comparison system and an overvaluing system, which are defined using vector norms [BOR 74].

6.1. Study of the stability of nonlinear systems from vector norms

6.1.1. *Vector norms*

Let $\mathrm{E} = \mathbb{R}^n$ be a vector space and $\mathrm{E}_1, \mathrm{E}_2 \ldots, \mathrm{E}_k$ be subspaces of E such that $\mathrm{E} = \mathrm{E}_1 \cup \mathrm{E}_2 \ldots \cup \mathrm{E}_k$ and let $\mathbf{x}$ be a vector defined on E with its projection on E_i denoted as $\mathbf{x}_i$.

$$\mathbf{x}_i = P_{r_i} \mathbf{x}, \qquad [6.1]$$

where P_{r_i} is the projection operator of E in E_i.

Let us note $p_i(\mathbf{x}) = p_i(\mathbf{x}_i)$, a scalar norm of vector $\mathbf{x}_i$ and $p(\mathbf{x}) = \left[p_1(\mathbf{x}), p_2(\mathbf{x}) \ldots, p_k \mathbf{x}_1 \right]^T$.

If $\mathbf{y}$ is a vector of E with $\mathbf{y}_i = P_{r_i}\mathbf{y}$, we obtain the following properties:

$$\begin{cases} p_i(\mathbf{x}_i) \geq 0, \forall \mathbf{x}_i \in \mathrm{E}_i \; \forall i = 1,2,\ldots,k \\ p_i(\mathbf{x}_i) = 0 \leftrightarrow \mathbf{x}_i = 0, \forall i = 1,2,\ldots,k \\ p_i(\mathbf{x}_i + \mathbf{y}_i) \leq p_i(\mathbf{x}_i) + p_i(\mathbf{y}_i), \forall \mathbf{x}_i, \mathbf{y}_i \in \mathrm{E}_i \; \forall i = 1,2,\ldots,k \\ p_i(\lambda \mathbf{x}_i) = |\lambda| p_i(\mathbf{x}_i), \forall \mathbf{x}_i \in \mathrm{E}_i, \forall i = 1,2,\ldots,k, \forall \lambda \cup \mathbb{R} \end{cases} \quad [6.2]$$

In what follows, subspaces E_i are disjointed and are all required in order to define the entire space E. Vector norm p is then said to be regular [BOR 87, BOR 08, BOR 74].

6.1.2. *Comparison systems and overvaluing systems*

A process comparison system is a system in which certain properties imply the same characteristics in the initial process. In studies on stability, the most frequently used comparison systems are overvaluing systems.

6.1.2.1. *Continuous case*

Let us consider process S of state vector $\mathbf{x} \in \mathbb{R}^n$, for which evolution is governed by equation:

$$\dot{\mathbf{x}} = \mathbf{A}(t, \mathbf{x}, \boldsymbol{\delta})\mathbf{x}, \quad t \in \tau = [t_0 + \infty], \quad \boldsymbol{\delta} \in \Delta = \{\boldsymbol{\delta}, \|\boldsymbol{\delta}\| \leq \boldsymbol{\delta}_m\} \quad [6.3]$$

where $\boldsymbol{\delta}$ represents a vector that groups model uncertainties.

$$\begin{aligned} &\mathbf{A} : \tau \times \mathbb{R}^n, \Delta \rightarrow \mathbb{R}^{n \times n} \\ &\mathbf{A}_{ij}(t, \mathbf{x}, \boldsymbol{\delta}) = P_{r_i}\mathbf{A}(t, \mathbf{x}, \boldsymbol{\delta})P_{r_i}, \quad \forall i, j = 1,2,\ldots,k \end{aligned} . \quad [6.4]$$

The matrix $\mathbf{M} : \tau \times \mathbb{R}^{nnx} \rightarrow \mathbb{R}^{k \times k}$ is a pseudo-overvaluing matrix of $\mathbf{A}$ relative to vector norm p if and only if the following inequality is verified along the trajectories of the system:

$$\frac{d}{dt} p(\mathbf{x}) \leq \mathbf{M}(t, \mathbf{x}) \, p(\mathbf{x}) \quad \forall (t, \mathbf{x}) \in \tau \times \mathbb{R}^n . \quad [6.5]$$

Let us note that

$$m_{ij}(t,\mathbf{x},\mathbf{y},\boldsymbol{\delta}) = \frac{\text{grad } p_i(\mathbf{y}_i)^T \mathbf{A}_{ij}(t,\mathbf{x},\boldsymbol{\delta})\mathbf{y}_i}{p_i(\mathbf{y}_i)} . \qquad [6.6]$$

DEFINITION 6.1.–

Matrix $\mathbf{M}(t,\mathbf{x}) = \{\mu_{ij}(t,\mathbf{x})\}$, $\mu_{ij}(t): \tau \times \mathbb{R}^n \to \mathbb{R}$,

$$\mu_{ii}(t) = \sup_{\substack{\mathbf{y}\in\mathbb{R}^n \\ \boldsymbol{\delta}\in\Delta}} m_{ii}(t,\mathbf{x},\mathbf{y},\boldsymbol{\delta}), \quad \forall i = 1,2,\ldots,k \qquad [6.7]$$

$$\mu_{ij}(t) = \sup_{\substack{\mathbf{y}\in\mathbb{R}^n \\ \boldsymbol{\delta}\in\Delta}} \left| m_{ij}(t,\mathbf{x},\mathbf{y},\boldsymbol{\delta}) \right|, \quad \forall i \neq j = 1,2,\ldots,k \qquad [6.8]$$

is a pseudo-overvaluing matrix of **A**.

DEFINITION 6.2.–

Matrix $\mathbf{M}(t) = \{\mu_{ij}(t)\}$, with

$$\mu_{ii}(t) = \sup_{\substack{\mathbf{y}\in\mathbb{R}^n \\ \boldsymbol{\delta}\in\Delta}} m_{ii}(t,\mathbf{y},\mathbf{y},\boldsymbol{\delta}), \quad \forall i = 1,2,\ldots,k \qquad [6.9]$$

$$\mu_{ij}(t) = \sup_{\substack{\mathbf{y}\in\mathbb{R}^n \\ \boldsymbol{\delta}\in\Delta}} \left| m_{ij}(t,\mathbf{y},\mathbf{y},\boldsymbol{\delta}) \right|, \quad \forall i \neq j = 1,2,\ldots,k \qquad [6.10]$$

is a pseudo-overvaluing matrix of **A**.

DEFINITION 6.3.–

The constant matrix $\mathbf{M} = \{\mu_{ij}\}$, with

$$\mu_{ii}(t) = \sup_{\substack{\mathbf{y}\in\mathbb{R}^n \\ t\in\tau \\ \boldsymbol{\delta}\in\Delta}} m_{ii}(t,\mathbf{y},\mathbf{y},\boldsymbol{\delta}), \quad \forall i = 1,2,\ldots,k \qquad [6.11]$$

$$\mu_{ij}(t) = \sup_{\substack{\mathbf{y}\in\mathbb{R}^n \\ t\in\tau \\ \boldsymbol{\delta}\in\Delta}} \left| m_{ij}(t, \mathbf{y}, \mathbf{y}, \boldsymbol{\delta}) \right|, \quad \forall i \neq j = 1, 2, \ldots, k \qquad [6.12]$$

is a pseudo-overvaluing matrix of **A**.

We obtain $\mathbf{M}(t,\mathbf{x}) \leq \mathbf{M}(t) \leq \mathbf{M}$.

With these definitions, the system for which evolution is governed by the equation:

$$\dot{\mathbf{z}} = \mathbf{M}(.)\mathbf{z} \qquad [6.13]$$

is a comparison system for the initial system and it verifies $p(\mathbf{x}(t)) < \mathbf{z}(t)$ $\forall t \in \tau$, if this property was verified at the initial instant:

$$p(\mathbf{x}(t_0)) < \mathbf{z}(t_0). \qquad [6.14]$$

The stability of the comparison system implies the stability of the initial system.

If $\mathbf{M}(.)$ is a pseudo-overvaluing matrix of **A**, the matrix $\mathbf{M}(.) + \Delta\mathbf{M}$ (where $\Delta\mathbf{M}$ is a matrix with all positive or zero elements) is also a pseudo-overvaluing matrix of **A**.

6.1.2.2. *Discrete case*

In this case, the process evolution is governed by an equation of the form:

$$\mathbf{x}(t+1) = \mathbf{A}(t, \mathbf{x}(t), \boldsymbol{\delta}(t))\mathbf{x}(t) \qquad [6.15]$$

where vector $\boldsymbol{\delta}$ includes all uncertainties and disturbances that act on the process:

$$\boldsymbol{\delta} \in \Delta = \left\{ \boldsymbol{\delta}, \|\boldsymbol{\delta}\| \leq \boldsymbol{\delta}_m \right\} \qquad [6.16]$$

The upper bound can be carried out similarly to the continuous case.

DEFINITION 6.4.–

Matrix $\mathbf{M}(t, \mathbf{x}(t)) = \left\{ \mu_{ij}(t, \mathbf{x}(t)) \right\}$, with

$$\mu_{ij}(t) = \sup_{\substack{\mathbf{y}\in\mathbb{R}^n \\ \boldsymbol{\delta}\in\Delta}} \left| m_{ij}(t, \mathbf{x}(t), \mathbf{y}, \boldsymbol{\delta}) \right|, \quad \forall i = 1, 2, \ldots, k \qquad [6.17]$$

is an overvaluing matrix of matrix **A**.

DEFINITION 6.5.–

Matrix $\mathbf{M}(t)=\{\mu_{ij}(t)\}$, with

$$\mu_{ij}(t)=\sup_{\substack{\mathbf{y}\in\mathbb{R}^n\\ \boldsymbol{\delta}\in\Delta}}\left|m_{ij}(t,\mathbf{y},\mathbf{y},\boldsymbol{\delta})\right|,\quad \forall i\neq j=1,2,\ldots,k \tag{6.18}$$

is an overvaluing matrix of matrix **A**.

DEFINITION 6.6.–

The constant matrix $\mathbf{M}=\{\mu_{ij}\}$,

$$\mu_{ij}(t)=\sup_{\substack{\mathbf{y}\in\mathbb{R}^n\\ t\in\tau\\ \boldsymbol{\delta}\in\Delta}}\left|m_{ij}(t,\mathbf{y},\mathbf{y},\boldsymbol{\delta})\right|,\quad \forall i,j=1,2,\ldots,k \tag{6.19}$$

is an overvaluing matrix of matrix **A**.

We obtain $\mathbf{M}(t,\mathbf{x})\leq\mathbf{M}(t)\leq\mathbf{M}$.

As in the continuous case, if $\mathbf{M}(.)$ is an overvaluing matrix of **A**, matrix $\mathbf{M}^{'}(.)=\mathbf{M}(.)+\Delta\mathbf{M}$, where $\Delta\mathbf{M}$ is a matrix with all positive or zero elements, is also an overvaluing of **A**.

With these definitions, the system for which evolution is governed by equation $\mathbf{z}(t+1)(.)=\mathbf{M}(.)\mathbf{z}(t)$ is a comparison system of the initial process, and the overvaluation $p(\mathbf{x}(t))<\mathbf{z}(t)\quad\forall t\in\tau$ is verified if this property is verified at the initial instant $p(\mathbf{x}(t_0))<\mathbf{z}(t_0)$.

Therefore, the stability of the comparison system implies that of the initial system.

REMARK 6.1.–

For a representation of process evolution by an equation of the form:

$$\dot{\mathbf{x}}(t)=\mathbf{A}(t,\mathbf{x}(t),\boldsymbol{\delta})\mathbf{x}(t)+\mathbf{B}(t,\mathbf{x}(t),\boldsymbol{\delta}) \tag{6.20}$$

or

$$\mathbf{x}(t+1) = \mathbf{A}(t, \mathbf{x}(t), \boldsymbol{\delta}(t))\mathbf{x}(t) + \mathbf{B}(t, \mathbf{x}(t), \boldsymbol{\delta}(t)) \qquad [6.21]$$

the overvaluation is carried out similarly to those presented above. We obtain:

$$\frac{d}{dt} p(\mathbf{x}) \leq \mathbf{M}(.)\, p(\mathbf{x}) + \mathbf{N}(.) \qquad [6.22]$$

or

$$p(\mathbf{x}(t+1)) \leq \mathbf{M}(.)\, p(\mathbf{x}(t)) + \mathbf{N}(.). \qquad [6.23]$$

In both cases, for $\mathbf{N}(.)$ we have:

$$\mathbf{N}(.) = \{\mathbf{v}_i\ (.)\} \qquad [6.24]$$

with

$$\mathbf{v}_i(t, \mathbf{x}) = \sup_{\boldsymbol{\delta} \in \Delta} p_i\left(\mathbf{B}(t, \mathbf{x}, \boldsymbol{\delta})\right) \qquad [6.25]$$

$$\mathbf{v}_i(t) = \sup_{\substack{\mathbf{x} \in \mathbb{R}^n \\ \boldsymbol{\delta} \in \Delta}} p_i\left(\mathbf{B}(t, \mathbf{x}, \boldsymbol{\delta})\right) \qquad [6.26]$$

$$\mathbf{v}_i = \sup_{\substack{\mathbf{x} \in \mathbb{R}^n \\ t \in \mathcal{T} \\ \boldsymbol{\delta} \in \Delta}} p_i\left(\mathbf{B}(t, \mathbf{x}, \boldsymbol{\delta})\right). \qquad [6.27]$$

REMARK 6.2.–

For $\mathbf{x} \in \mathbb{R}^n$, the stability conditions obtained from the comparison system relative to vector norm p are less conservative if vector norm $p(\mathbf{x})$ is itself of size n.

In this case, the simplest vector norm is written as:

$$p(\mathbf{x}) = |P\mathbf{x}|$$

P being a reversible square matrix $n \times n$ if $\mathbf{v} = [\mathbf{v}_1, \mathbf{v}_2, \ldots, \mathbf{v}_n]^T$,

$$|\mathbf{v}| = [|\mathbf{v}_1|, |\mathbf{v}_2|, \ldots, |\mathbf{v}_n|]^T.$$

For $P = I$, we obtain:

$$p(\mathbf{x}) = |\mathbf{x}| = [|\mathbf{x}_1|, |\mathbf{x}_2| \ldots, |\mathbf{x}_n|]^T. \qquad [6.28]$$

For P different from the identity matrix, the method means taking $p(\mathbf{x}) = |\mathbf{x}'|$ with the base change corresponding to $\mathbf{x}' = P\mathbf{x}$.

When $p(\mathbf{x}) = |\mathbf{x}|$, the search for a comparison system is simplified for $\mathbf{A}(.) = \{a_{ij}(.)\}$ and $\mathbf{B}(.) = \{b_{ij}(.)\}$.

We obtain:

1) Continuous case

$$\begin{aligned} m_{ii}(.) &= a_{ii}(.), & \forall i &= 1, \ldots, n \\ m_{ij}(.) &= |a_{ij}(.)|, & \forall i \neq j &= 1, \ldots, n \\ n_i(.) &= |b_i(.)|, & \forall i &= 1, \ldots, n \end{aligned} . \qquad [6.29]$$

2) Discrete case

$$\begin{aligned} m_{ij}(.) &= |a_{ij}(.)|, & \forall i \neq j &= 1, \ldots, n \\ n_i(.) &= |b_i(.)|, & \forall i &= 1, \ldots, n \end{aligned} . \qquad [6.30]$$

The terms $\mu_{ij}(.)$ and $\mathbf{v}_i(.)$ can be directly deduced.

6.1.3. *Determination of attractors*

For stable homogeneous system:

$$\begin{aligned} \dot{\mathbf{x}} &= \mathbf{A}(t, \mathbf{x}, \boldsymbol{\delta})\mathbf{x} \\ \mathbf{x}(t+1) &= \mathbf{A}(t, \mathbf{x}(t), \boldsymbol{\delta})\mathbf{x}(t) \end{aligned} \qquad [6.31]$$

The attractor is reduced at the origin.

For a locally unstable non-homogeneous system, the attractor is an area surrounding the origin.

When the process comparison system is constant, the attractor can be determined quite simply.

6.1.3.1. *Continuous case [GHA 13]*

$$\dot{\mathbf{z}} = \mathbf{M}\mathbf{z} + \mathbf{N} \qquad [6.32]$$

If the matrix **M** with negative diagonal elements and positive or zero off-diagonal elements is the opposite of an M-matrix, meaning that all its eigenvalues are stable in the continuous meaning (with strictly negative real parts), we get:

$$\mathbf{z}_\infty = \lim_{t\to\infty} \mathbf{z}(t) = -\mathbf{M}^{-1}\mathbf{N} \qquad [6.33]$$

and under such conditions as $p(\mathbf{x}(t)) \le \mathbf{z}(t)$

$$\lim_{t\to\infty} p(\mathbf{x}(t)) \le \mathbf{z}_\infty = -\mathbf{M}^{-1}\mathbf{N} \qquad [6.34]$$

the attractor corresponds to a domain $D = \left\{ \mathbf{x} \in \mathbb{R}^n ; p(\mathbf{x}) \le -\mathbf{M}^{-1}\mathbf{N} \right\}$.

The condition that **M** be the opposite of an M-matrix is expressed simply from the Kotelyanski conditions $\mathbf{M} = \left\{ m_{ij} \right\}$.

$$m_{1,1} \prec 0, \begin{vmatrix} m_{1,1} & |m_{1,2}| \\ |m_{2,1}| & m_{2,2} \end{vmatrix} \succ 0, \ldots, (-1)^k \begin{vmatrix} m_{1,1} & |m_{1,2}| & \cdots & |m_{1,k}| \\ |m_{2,1}| & m_{2,2} & \cdots & |m_{2,k}| \\ \vdots & \vdots & & \vdots \\ |m_{k,1}| & |m_{k,2}| & \cdots & m_{k,k} \end{vmatrix} \succ 0 \cdot \qquad [6.35]$$

This means that the determinants of successive principal minors of M must be of alternating signs with the first being negative.

6.1.3.2. *Discrete case [BOR 76]*

$$\mathbf{z}(t+1) = \mathbf{M}\mathbf{z}(t) + \mathbf{N} \ . \qquad [6.36]$$

If matrix **M** with all positive elements has module eigenvalues that are all strictly lower than a single unit, for t tending to infinity we obtain:

$$\mathbf{z}(t) \rightarrow \mathbf{z}_{\infty} = (I - \mathbf{M})^{-1}\,\mathbf{N} \qquad [6.37]$$

where I is the identity matrix.

The condition for the spectral radius of M to be of lower module than the unit is simply verified by imposing the Kotelyanski conditions on the matrix $(I - \mathbf{M})$, which is then the opposite of an M-matrix.

Attractor D is then defined by:

$$D = \left\{\mathbf{x} \,/\, p(\mathbf{x}) \leq \mathbf{z}_{\infty} = (I - \mathbf{M})^{-1}\,\mathbf{N}\right\} \qquad [6.38]$$

REMARK 6.3.–

In continuous and discrete cases, the least conservative conditions are obtained if the vector norm $p(\mathbf{x})$ is of size n.

REMARK 6.4.–

The result can be further improved if the system matrix $\mathbf{A}(.)$ can be linearized near the origin with real and distinct eigenvalues. For transition matrix P_D in $p(\mathbf{x}) = |P_D\mathbf{x}|$, we then just need to take a matrix that diagonalizes the linear $\mathbf{A}_L$ of $\mathbf{A}(.)$ near the origin.

If this is not possible, the result can still be improved by taking $p(\mathbf{x}) = |P_F\mathbf{x}|$, where P_F is the base change that puts the linear of $\mathbf{A}(.)$ into an arrow form (Appendix A), with the diagonal elements of P_F having been selected to minimize the modules of elements from the last row and the last column of $P_F^{-1}\mathbf{A}_L P_F$.

REMARK 6.5.–

In the search for an estimate by overvaluation of the error in the case of trajectory tracking, we obtain:

$$e(t) = \mathbf{x}_R(t) - \mathbf{x}(t),$$

where $\mathbf{x}_R(t)$ represents the reference state trajectory.

6.1.4. *Nested attractors [GHA 15a]*

Having determined a first attractor D_1 from an initial chosen representation, it is possible to choose another representation for $\mathbf{x} \in D_1$

$$\begin{aligned}&\dot{\mathbf{x}} = \mathbf{A}^{'}(t, \mathbf{x}, p)\mathbf{x} + \mathbf{B}^{'}(t, \mathbf{x}, p)\\&\mathbf{x}(t+1) = \mathbf{A}^{'}(t+1, \mathbf{x}(t), p(t))\mathbf{x}(t) + \mathbf{B}^{'}(t, \mathbf{x}(t), p(t))\end{aligned} \quad [6.39]$$

and to start the overvaluation again, this time with $\mathbf{x} \in D_1$, not $\mathbf{x} \in \mathbb{R}^n$.

This operation can be repeated several times, allowing us to define nested attractors $D_1 \supset D_2 \supset D_3 \dots$.

6.2. Adaptation of control

6.2.1. *Minimizing the size of attractors: direct approach*

Process command law depends on the parameter $\theta \in \mathbb{R}^2$.

We obtain the representation:

$$\begin{aligned}&\dot{\mathbf{x}}(t) = \mathbf{A}(t, \mathbf{x}, u(t, \mathbf{x}, \theta), p)\mathbf{x}(t) + \mathbf{B}(.)\\&\mathbf{x}(t+1) = \mathbf{A}(t, \mathbf{x}(t), u(t, \mathbf{x}(t), \theta), p(t))\mathbf{x}(t) + \mathbf{B}(.)\end{aligned} \quad [6.40]$$

that are expressions in which the vector $\mathbf{B}(.)$ is module limited.

In this case, determining a linear comparison system results in the following expression:

$$\begin{aligned}&\dot{\mathbf{z}} = \mathbf{M}(\theta)\mathbf{z}(t) + \mathbf{N}(\theta)\\&\mathbf{z}(t+1) = \mathbf{M}(\theta)\mathbf{z}(t) + \mathbf{N}(\theta)\end{aligned}. \quad [6.41]$$

If matrix $\mathbf{M}(\theta)$ is stable, in the continuous case we get:

$$\lim_{t \to \infty} p(\mathbf{x}(t)) \leq -\mathbf{M}^{-1}(\theta)\mathbf{N}(\theta) = p_{\infty}(\theta) \quad [6.42]$$

and in the discrete case:

$$\lim_{t\to\infty} p(\mathbf{x}(t)) \leq \left(I - \mathbf{M}^{-1}\right)(\theta)\mathbf{N}(\theta) = p_\infty(\theta)\,. \qquad [6.43]$$

In the domain of evolution of θ where the matrix $\mathbf{M}(\theta)$ is stable, it is therefore necessary to look for the value of θ that minimizes a norm p_∞ by using optimization techniques [BOR 13].

6.2.2. *Minimizing the size of attractors by metaheuristics*

This approach is similar to the previous one, the only difference is the algorithm implemented for optimization.

Indeed, having defined the attractor through a relationship of the form $p(\mathbf{x}) \leq p_\infty(\theta)$, the $p_\infty(\theta)$ term can be particularly complex and it is often difficult to apply conventional optimization techniques that were mostly derived from the gradient method. In this case, a simple implementation of metaheuristics, such as tabu search, is particularly efficient [STE 14, GHA 16].

6.3. Overvaluation of the maximum error for various applications

6.3.1. *Control of nonlinear systems by pole placement*

1) *Continuous case* [BOR 90, GHA 16b]

The evolution of the process is described by equation:

$$\dot{\mathbf{x}} = \mathbf{A}(.)\mathbf{x}(t)+\mathbf{B}(.)u(t) + \mathbf{B}'(.)\,. \qquad [6.44]$$

All uncertainties and disturbances are grouped in vector $\mathbf{B}'(.)$ with bounded evolution:

$$\mathbf{B}'_m \leq \mathbf{B}'(.) \leq \mathbf{B}'_M\,. \qquad [6.45]$$

The linearized process near the origin can be written as:

$$\dot{\mathbf{x}}_l(t) = \mathbf{A}(0)\mathbf{x}_l+\mathbf{B}(0)u\,. \qquad [6.46]$$

Pole placement is done through state feedback such as:

$$u = -\mathbf{L}\mathbf{x}_l \qquad [6.47]$$

where vector **L** allows us to impose the selected poles.

This state feedback, calculated from the linearized model, is applied to the initial process. We obtain:

$$\dot{\mathbf{x}} = \mathbf{A}(\mathbf{x})\mathbf{x}+\mathbf{B}(\mathbf{x})\mathbf{L}\mathbf{x} + \mathbf{B}^{'}(.) \qquad [6.48]$$

$$\dot{\mathbf{x}}(t) = \mathcal{A}(\mathbf{x}(t))\mathbf{x}(t)+\mathbf{B}^{'}(.) \qquad [6.49]$$

with

$$\mathcal{A}(\mathbf{x}(t)) = \mathbf{A}(\mathbf{x}) + \mathbf{B}(\mathbf{x})\mathbf{L}\,. \qquad [6.50]$$

Determining the comparison system is done directly from the closed-loop system.

A very detailed implementation of the method on an academic example of the third order is presented in Appendix B [GHA 13].

2) *Discrete case*

The approach is similar to the continuous case:

$$\dot{\mathbf{x}}(t+1) = \mathbf{A}(.)\mathbf{x}(t)+\mathbf{B}(.)u(t) + \mathbf{B}^{'}(.)\,. \qquad [6.51]$$

Uncertainties and disturbances are grouped from vector $\mathbf{B}^{'}(.)$ with bounded evolution.

$$\mathbf{B}^{'}_{m} \leq \mathbf{B}^{'}(.) \leq \mathbf{B}^{'}_{M}\,. \qquad [6.52]$$

Pole placement is done through state feedback such as:

$$u(t) = -\mathbf{L}\mathbf{x}_l(t) \qquad [6.53]$$

calculated from the linearized model:

$$\dot{\mathbf{x}}_e(t+1) = \mathbf{A}_e\mathbf{x} + \mathbf{x}_e(t)+\mathbf{B}_e u(t) \qquad [6.54]$$

of the initial system.

Determining the comparison system for identifying an upper bound of the error is done from the initial system where state feedback is calculated on the linearized model.

6.3.2. *Diffeomorphism command of nonlinear processes*

For a nonlinear multivariable process, prior to determining the command, it may be of benefit to carry out a process decoupling linearization.

Input–output linearization through state feedback is an approach that has attracted much research in recent years [JAK 80, HED 05, LEV 04, MAR 93, BOR 90].

Let us consider the supposed controllable process for which evolution is described by a model of the form:

$$\begin{aligned} \dot{\mathbf{x}} &= \mathbf{f}(\mathbf{x}) + \mathbf{G}(\mathbf{x})u \\ \mathbf{y} &= \mathbf{h}(\mathbf{x}) \end{aligned} \qquad [6.55]$$

with: $\mathbf{x} = [\mathbf{x}_1, \mathbf{x}_2, \ldots, \mathbf{x}_m]^T$, $\mathbf{y} = [\mathbf{y}_1, \mathbf{y}_2, \ldots, \mathbf{y}_m]^T$ and $u = [u_1, u_2, \ldots, u_m]^T$.

The method involves deriving each component $\mathbf{y}_i$ of the output vector until the command vector appears.

We note:

$$\begin{aligned} &\mathbf{h}(\mathbf{x}) = [\mathbf{h}_1(\mathbf{x}), \quad \mathbf{h}_2(\mathbf{x}), \quad \cdots, \quad \mathbf{h}_m(\mathbf{x})]^T \\ &\mathbf{h}_i(\mathbf{x}) = \mathbf{h}_i^0(\mathbf{x}) \\ &\mathbf{h}_i^j(\mathbf{x}) = \mathbf{h}_{i,\mathbf{x}}^{j-1T}(\mathbf{x})\mathbf{f}(\mathbf{x}) \\ &\mathbf{h}_{i,\mathbf{x}}^j = \frac{\partial \mathbf{h}_i^j(\mathbf{x})}{\partial \mathbf{x}} \\ &\mathbf{G}_i^j(\mathbf{x}) = h_{i,\mathbf{x}}^{j-1T}(\mathbf{x})\mathbf{G}(\mathbf{x}) \end{aligned} \qquad [6.56]$$

with

$$\forall j \le d_i, \ \mathbf{G}_i^j(\mathbf{x}) = 0 \ \text{ and } \ \mathbf{G}_i^{d_i+1}(\mathbf{x}) \ne 0 \qquad [6.57]$$

we obtain

$$\mathbf{y}^* = \begin{pmatrix} \mathbf{y}_1^{(d_1+1)} \\ \vdots \\ \mathbf{y}_i^{(d_i+1)} \\ \vdots \\ \mathbf{y}_m^{(d_m+1)} \end{pmatrix}, \; \mathbf{f}^* = \begin{pmatrix} \mathbf{h}_1^{(d_1+1)}(\mathbf{x}) \\ \vdots \\ \mathbf{h}_i^{(d_i+1)}(\mathbf{x}) \\ \vdots \\ \mathbf{h}_m^{(d_m+1)}(\mathbf{x}) \end{pmatrix}, \; \mathbf{f}^* = \begin{pmatrix} \mathbf{G}_1^{(d_1+1)}(\mathbf{x}) \\ \vdots \\ \mathbf{G}_i^{(d_i+1)}(\mathbf{x}) \\ \vdots \\ \mathbf{G}_m^{(d_m+1)}(\mathbf{x}) \end{pmatrix}, \text{ and } \mathbf{y}^* = \mathbf{f}^* + \mathbf{G}^* u . \quad [6.58]$$

If $\mathbf{G}^*$ is reversible in the area of evolution of the state vector, by noting:

$$u = \mathbf{G}^{*-1}(\mathbf{x})(\mathbf{v} - \mathbf{h}^*(\mathbf{x})) \quad [6.59]$$

$$\mathbf{y}_i^{(d_i+1)} = \mathbf{v}_i, \forall i = 1, 2, \ldots, m \quad [6.60]$$

we get Figure 6.1.

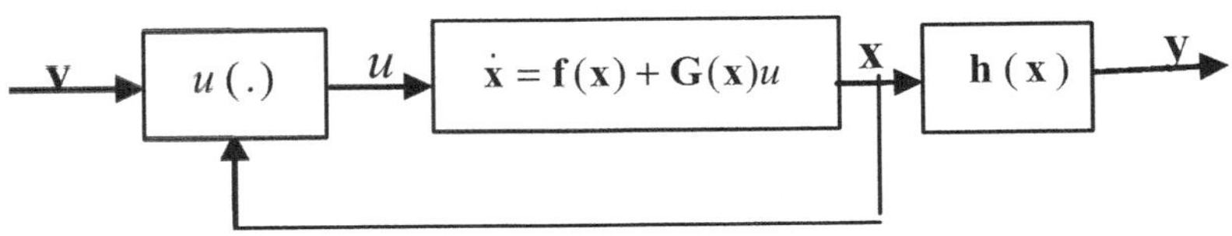

Figure 6.1. *Block diagram of input–output linearization by state feedback*

This representation is only valid if the state feedback has not made the unstable elements of the process unobservable.

This state feedback transforms each decoupled subsystem into a chain of integrators (Figure 6.2).

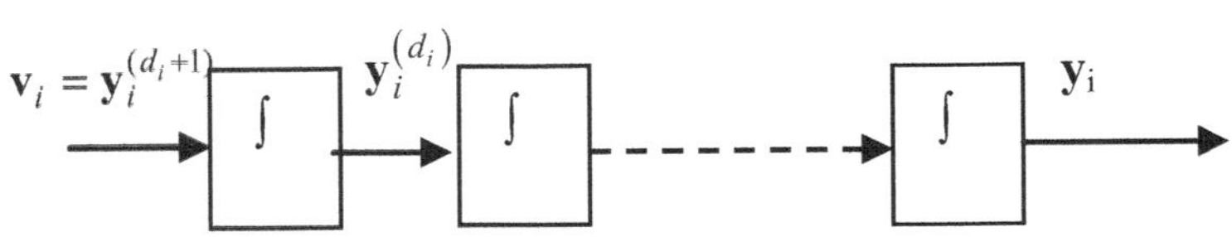

Figure 6.2. *Decoupled subsystem in a chain of integrators*

By replacing u and $\mathbf{v}$ by their values, we obtain, for $\mathbf{y}^c = 0$, a description of the form:

$$\frac{d\mathbf{x}}{dt} = \mathbf{A}(\mathbf{x},t)\mathbf{x}+\mathbf{B}(\mathbf{x},t)\,. \quad [6.61]$$

To investigate the stability of the system, we need to determine the attractor of the system. This attractor should also possibly be reduced to the origin $x = 0$ by using aggregation techniques and developing a comparison system such that its instantaneous characteristic matrix is the opposite of an M-matrix.

For the vector norm $p(\mathbf{x}) = \left[|\mathbf{x}_1|,|\mathbf{x}_2|\ldots,|\mathbf{x}_n|\right]^T$, by upper bound we get the following comparison system:

$$\mathbf{z} \in \mathbb{R}^n / \dot{\mathbf{z}}(t) = \mathbf{M}\mathbf{z}(t) + \mathbf{N} \quad [6.62]$$

where **M** and **N** are matrices with constant coefficients.

If $p(\mathbf{x}_0) \leq \mathbf{z}_0$ at initial instant t_0, we get $p(\mathbf{x}) \leq \mathbf{z}, \forall t > 0$.

Attractor D is then defined by: $D = \left\{\mathbf{x}/; p(\mathbf{x}) \leq -\mathbf{M}^{-1}\mathbf{N}\right\}$.

An academic example of application for a second-order system is presented in detail in Appendix C [GHA 15].

6.3.3. *Determining the attractor for Lur'e Postnikov type processes [GHA 14]*

Models with separable nonlinearity from Lur'e Postnikov type models are widely used in process control, which justifies our interest in this class of processes.

6.3.3.1. *Description of studied systems*

Let us consider the Lur'e Postnikov type of system described in Figure 6.3.

The linear part is characterized by the matrix triplet (**A**, **B**, **C**) and the nonlinearity is defined in a domain limited by the lines $f_1(\varepsilon)=f_M\varepsilon$ and $f_2(\varepsilon)=f_m\varepsilon$ (Figure 6.4).

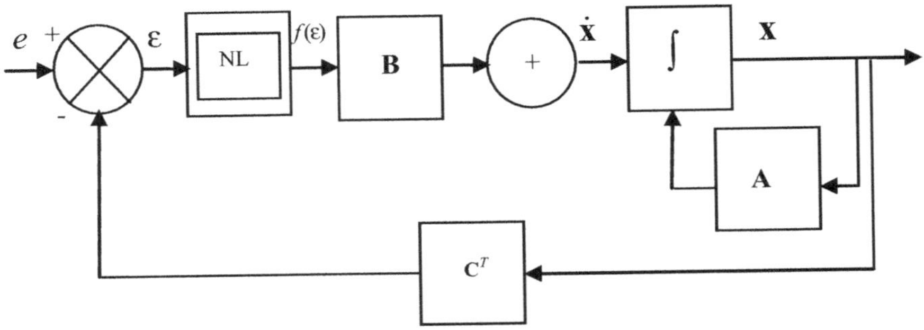

Figure 6.3. *Lur'e Postnikov type process*

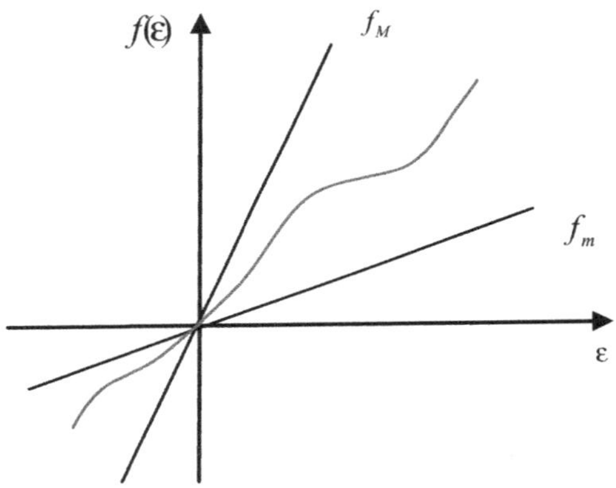

Figure 6.4. *Nonlinearity of the studied system*

If $\left.\frac{f(\varepsilon)}{\varepsilon}\right|_{\varepsilon=0}$ is defined, then $\frac{f(\varepsilon)}{\varepsilon}$, denoted $f^{*}(\varepsilon)$, is assumed such that $f_m \leq f^{*}(\varepsilon) \leq f_M$.

$$\begin{cases} \dot{\mathbf{x}} = \mathbf{Ax}+\mathbf{B}f(\varepsilon) \\ \varepsilon = e - \mathbf{C}^T\mathbf{x} \end{cases} \qquad [6.63]$$

or even

$$\dot{\mathbf{x}} = \left(\mathbf{A}\text{-}\mathbf{BC}^T f^{*}(.)\right)\mathbf{x}+\mathbf{B}f^{*}(.)e . \qquad [6.64]$$

In an autonomous system ($e=0$), the system is assumed to be subjected to disturbance $\delta(.)$ and its evolution is described by the equation:

$$\dot{\mathbf{x}} = \psi(\mathbf{x})\mathbf{x}+\delta(.) \qquad [6.65]$$

with

$$\psi(.) = \mathbf{A}\text{-}\mathbf{B}\mathbf{C}^T f^*(.) \ . \qquad [6.66]$$

6.3.3.2. *Case of a non-autonomous uncertain system (e is non-zero)*

Let us consider the following systems [6.63] and [6.64]:

$$\begin{cases} \dot{\mathbf{x}}_1 = \mathbf{A}\mathbf{x}_1 + \mathbf{B}f(\varepsilon_1) \\ \varepsilon_1 = e - \mathbf{C}^T\mathbf{x}_1 \end{cases} \qquad [6.67]$$

$$\begin{cases} \dot{\mathbf{x}}_2 = \mathbf{A}\mathbf{x}_1 + \mathbf{B}f(\varepsilon_2) + \delta(.) \\ \varepsilon_2 = e - \mathbf{C}^T\mathbf{x}_2 \end{cases} , \qquad [6.68]$$

where [6.67] corresponds to the undisturbed system and [6.68] to the disturbed system with:

– $\mathbf{x}_1$ desired trajectory, meaning that of the undisturbed system;

– $\mathbf{x}_2$ disturbed trajectory;

– e process input set for $\mathbb{R}$;

– $\mathbf{A}$ constant square matrix $n \times n$;

– $\mathbf{B}$ constant vector of order n of elements b_i , $\forall i = 1,\ldots,n$;

– $\mathbf{C}$ constant vector of order n;

– $\delta(.)$ disturbance.

The evolution of the error between these systems can be described by the following equation:

$$\begin{cases} \dot{\mathbf{x}}_1\text{-}\dot{\mathbf{x}}_2 = \mathbf{A}\left(\mathbf{x}_1\text{-}\mathbf{x}_2\right) + \mathbf{B}\left(f(\varepsilon_1) - f(\varepsilon_2)\right) + \delta(.) \\ \varepsilon_1 - \varepsilon_2 = -\mathbf{C}^T\left(\mathbf{x}_1\text{-}\mathbf{x}_2\right) \end{cases} . \qquad [6.69]$$

Applying the mean value theorem gives:

$$f(\varepsilon_1)-f(\varepsilon_2)=f^*(.)(\varepsilon_1-\varepsilon_2) \quad [6.70]$$

$f^*(.)$ obtains its values among those of the derivative of f relative to ε of the nonlinear characteristics.

Let $\mathbf{x}_1\text{-}\mathbf{x}_2=\xi$, we obtain:

$$\dot{\xi}=\varphi(\mathbf{x})\xi-\delta(.) \quad [6.71]$$

with

$$\varphi(.)=\mathbf{A}\text{-}\mathbf{B}\mathbf{C}^T f^*(.) \quad [6.72]$$

– Determining the attractor or the deviation limit relative to the nominal trajectory by aggregation techniques

For the vector norm $p(\xi)=\left[|\xi_1|,|\xi_2|\ldots,|\xi_n|\right]^T$, we obtain the following comparison system:

$$\mathbf{z}\in\mathbb{R}^n/\dot{\mathbf{z}}(t)=\mathbf{M}(.)\mathbf{z}(t)+\mathbf{N} \quad [6.73]$$

with

$$p(\xi)\le\mathbf{z}\,. \quad [6.74]$$

If **M** and **N** are constant matrices and if **M** is the opposite of an M-matrix, as **N** is a positive vector, there then exists an asymptotically stable attractor D for the error ξ such as:

$$D=\left\{\xi\in\mathbb{R}^n;\,p(\xi)\le-\mathbf{M}^{-1}\mathbf{N}\right\}. \quad [6.75]$$

M is the matrix obtained by taking the maximum value of diagonal elements, the least favorable values of off-diagonal element modules of $\varphi(\mathbf{x})$ and **N** elements equal to the maxima of the modules of the elements of δ such that:

$$\mathbf{N}=\max\left|\boldsymbol{\delta}\right|. \quad [6.76]$$

If **M** is the opposite of an M-matrix, we get:

$$\lim_{t\to\infty} p(\xi) \leq \lim_{t\to\infty} \mathbf{z} = -\mathbf{M}^{-1}\mathbf{N}\ . \qquad [6.77]$$

This approach is presented in detail in Appendix D.

6.3.4. *Minimizing the attractor through tabu search*

The principle of tabu search [STE 14, GHE 07] involves starting from a solution in order to find the best solution in its neighborhood and using this new solution to start over; this is the principle of the greedy algorithm, which can quickly fall into a local optimum. The problem is avoided in the tabu search by forbidding the algorithm from falling in n_r recent solutions that are tabu, thereby allowing it to leave a local optimum by accepting a temporary deterioration of the solution found.

The following example illustrates this approach [GHA 16a].

Let us consider the nonlinear process of second order for which evolution is described by the equations:

$$\begin{cases} \dot{\mathbf{x}} = \mathbf{A}(\mathbf{x},t)\mathbf{x} + \mathbf{B}(x)u(x,\theta) + \delta(.) \\ \mathbf{y} = \mathbf{h}(\mathbf{x}) \end{cases} \qquad [6.78]$$

with

$$u(\theta,\mathbf{x}) = -\left(\theta_1\mathbf{y} + \theta_2\mathbf{x}_2\right) \qquad [6.79]$$

and

$$\mathbf{h}(\mathbf{x}) = \mathbf{x}_1 + (1 - e^{-\mathbf{x}_1^2})\mathbf{x}_2 \qquad [6.80]$$

with $\mathbf{x}(t) \in \mathbb{R}^2, \mathbf{B}(.) \in \mathbb{R}^2, \mathbf{A}(.) \in \mathbb{R}^{2\times2}, \theta \in \mathbb{R}^2$

$$\mathbf{A}(\mathbf{x},t) = \begin{bmatrix} -2 + \cos t + \cos \mathbf{x}_1 & 4 - e^{-\mathbf{x}_1^2}(1 + \sin \mathbf{x}_1) \\ 4 + \cos \mathbf{x}_2 & -8 + \sin \mathbf{x}_1 + e^{-\mathbf{x}_1^2} \end{bmatrix} \qquad [6.81]$$

$$\mathbf{B}(\mathbf{x}) = \begin{bmatrix} 3 + 0.5\cos \mathbf{x}_1 \\ 2 \end{bmatrix} \qquad [6.82]$$

we obtain

$$\dot{\mathbf{x}} = \mathcal{A}(\mathbf{x},t,\theta)\mathbf{x} + \boldsymbol{\delta}(.) \qquad [6.83]$$

with

$$\mathcal{A}(\mathbf{x},t,\theta) = \mathbf{A}(\mathbf{x},t) - \mathbf{B}(\mathbf{x})[\theta_1, \theta_1((1-e^{-\mathbf{x}_1^2})) + \theta_2] \qquad [6.84]$$

such that

$$\mathcal{A}(\mathbf{x},t,\theta) = \begin{bmatrix} a_{11} & a_{12} \\ a_{21} & a_{22} \end{bmatrix}$$

with

$$\begin{aligned} a_{11} &= -2 + \cos t \cos \mathbf{x}_1 - \theta_1(3 + 0.5\cos \mathbf{x}_1) \\ a_{12} &= 4 - e^{\mathbf{x}_1^2}(1 + \sin \mathbf{x}_1) - (3 + 0.5\cos \mathbf{x}_1)(\theta_1(1 - e^{\mathbf{x}_1^2}) + \theta_2) \\ a_{21} &= 4 + \cos \mathbf{x}_2 - 2\theta_1 \\ a_{22} &= -8 + e^{\mathbf{x}_1^2} + \sin \mathbf{x}_1 - 2[\theta_1(1 - e^{\mathbf{x}_1^2}) + \theta_2] \end{aligned} . \qquad [6.85]$$

Using the vector norm $p(\mathbf{x}) = [|\mathbf{x}_1| \quad |\mathbf{x}_2|]$, by overvaluation we obtain:

$$\frac{d}{dt} p(\mathbf{x}) \leq \mathbf{M}\left(\mathcal{A}(\mathbf{x},\theta,.)\right) p\left(\mathbf{x}\right) + \mathbf{N}(.) \qquad [6.86]$$

$$\mathbf{z} \in \mathbb{R}^n / \dot{\mathbf{z}}(t) = \mathbf{M}(.)\mathbf{z}(t) + \mathbf{N}(.) \qquad [6.87]$$

with

$$\mathbf{M}(\mathcal{A}(.)) = \begin{bmatrix} a_{11} & |a_{12}| \\ |a_{21}| & a_{21} \end{bmatrix} = \mathbf{M}(\mathbf{x},t,\theta_1,\theta_2) \qquad [6.88]$$

and

$$|\mathbf{N}(.)| \leq \boldsymbol{\delta}_M \qquad [6.89]$$

$\delta(.)$ is limited according to the following equations:

$$\boldsymbol{\delta}_1 = \begin{bmatrix} -0.2 \\ 0.3 \end{bmatrix} \leq \boldsymbol{\delta}(.) \leq \boldsymbol{\delta}_2 = \begin{bmatrix} 0.1 \\ 0.5 \end{bmatrix} \quad [6.90]$$

hence

$$\boldsymbol{\delta}_M = \begin{bmatrix} 0.2 \\ 0.5 \end{bmatrix}. \quad [6.91]$$

Starting from solution $\theta_1 = 2, \theta_2 = 0.1$, the possible variations at each stage are $\delta\theta_1$ and $\delta\theta_2$ with $|\delta\theta_1| = 0.2$ and $|\delta\theta_2| = 0.1$.

We obtain

$$\mathbf{M}(\mathbf{x}, t, 2, 0) = \begin{bmatrix} -8 + \cos t & \left| \begin{matrix} -2 - e^{-\mathbf{x}_1^2}(-5 + \sin \mathbf{x}_1 - \cos \mathbf{x}_1) \\ -\cos \mathbf{x}_1 \end{matrix} \right| \\ |\cos \mathbf{x}_2| & -12 + \sin \mathbf{x}_1 + 5e^{-\mathbf{x}_1^2} \end{bmatrix} \quad [6.92]$$

$$\mathbf{N} = \begin{bmatrix} 0.2 \\ 0.5 \end{bmatrix} \quad [6.93]$$

hence the comparison system

$$\dot{\mathbf{z}} = \begin{bmatrix} -7 & 3 \\ 1 & -7 \end{bmatrix} \mathbf{z} + \begin{bmatrix} 0.2 \\ 0.5 \end{bmatrix} \quad [6.94]$$

with the conditions

$$\begin{cases} -7 < 0 \\ \left(-1^2\right) \det(\mathbf{M}) > 0 \end{cases} \quad [6.95]$$

being verified if **M** is the opposite of an M-matrix. We obtain

$$p(\mathbf{x}) \leq -\mathbf{M}^{-1}\mathbf{N} = \begin{bmatrix} 0.0630 \\ 0.0804 \end{bmatrix} = p_\infty(\mathbf{x}). \quad [6.96]$$

Let us note

$$P_M = \|p_\infty(\mathbf{x})\| . \quad [6.97]$$

The strategy then involves searching in the vicinity of $\theta = [2;0]$ for the best solution in view of permissible variations. In what follows, the last three solutions found are tabu.

Starting with $\theta = [2;0]$, eight solutions are possible:

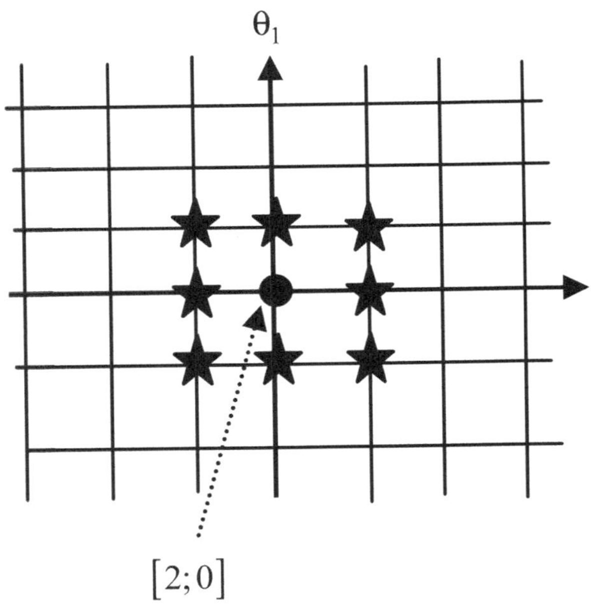

Figure 6.5. *Solutions near* $\theta = [2,0]$

$$\theta = (\theta_1, \theta_2 + \delta\theta_2), \theta = (\theta_1, \theta_2 - \delta\theta_2), \theta = (\theta_1 + \delta\theta_1, \theta_2), \theta = (\theta_1 - \delta\theta_1, \theta_2),$$
$$\theta = (\theta_1 + \delta\theta_1, \theta_2 + \delta\theta_2), \theta = (\theta_1 + \delta\theta_1, \theta_2 - \delta\theta_2), \theta = (\theta_1 - \delta\theta_1, \theta_2 + \delta\theta_2), \theta = (\theta_1 - \delta\theta_1, \theta_2 - \delta\theta_2)$$

The solution that minimizes the Euclidean norm of $p_\infty(\mathbf{x})$ matches $\theta=[2,2;0,1]$ and this solution becomes tabu. Continuing the iteration, the best solutions successively match $\theta=[2,4;0,2]$, $\theta=[2,4;0,3]$, $\theta=[2,4;0,4]$, $\theta=[2,4;0,5]$.

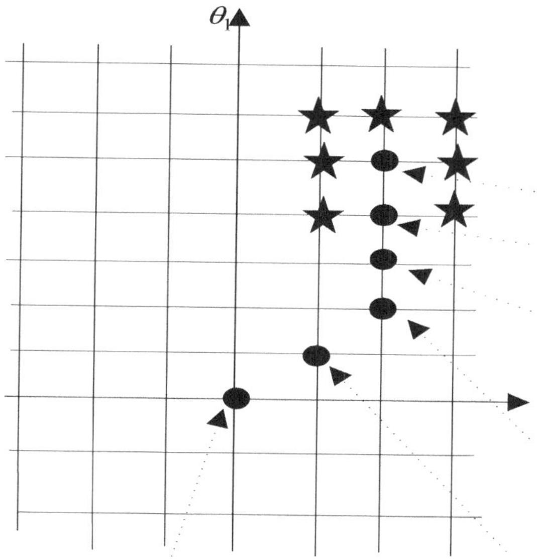

Figure 6.6. *Evolution of the best solutions*

From the latter solution, there cannot be any better improvement.

$$P_{\theta=[2,4:0,4]} = P_{\theta=[2,4:0,5]} = P_{\theta=[2,4:0,7]} = 0.870 \quad [6.98]$$

The best solution is therefore $\theta=[2,4;0,2]$.

The evolution of the state vector toward an attractor is shown in Figures 6.7 and 6.8.

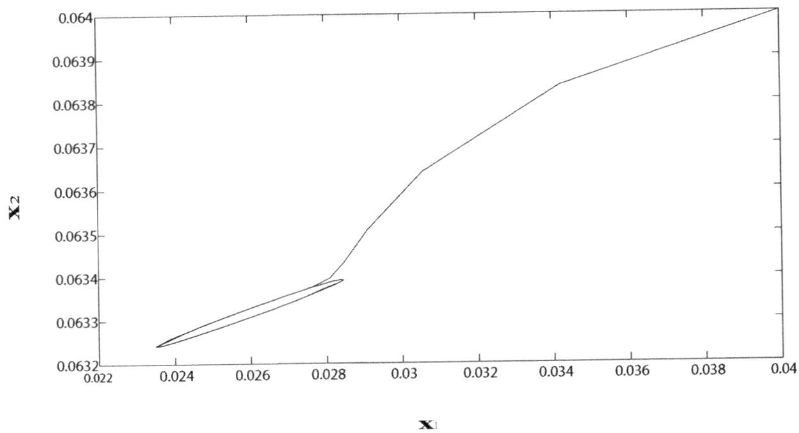

Figure 6.7. *Evolution of the state vector*

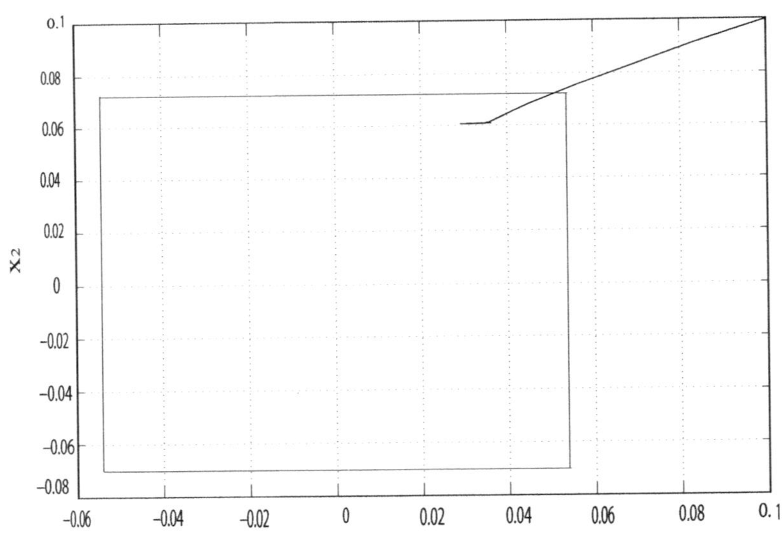

Figure 6.8. *Evolution of the state vector toward an attractor*

6.4. Fuzzy secondary loop control

Let us consider a nonlinear process that can be disturbed for which the command law C_M comes from a given model *M*. When the accuracy of instruction tracking is insufficient, the result can be improved by introducing a Ramdani fuzzy type secondary loop (Figure 6.9).

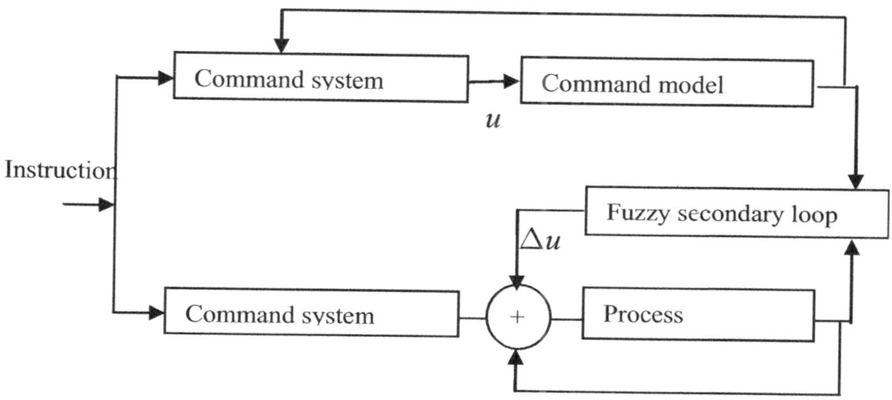

Figure 6.9. *Secondary regulation loop of fuzzy nature*

The secondary loop measures the difference between the process output and that of the model used to develop the command and derives the correction to be imposed on the command in order to best achieve the desired behavior.

7

Modeling and Control of an Elementary Industrial Process

In this chapter, our aim is to present the design of closed-loop control systems that are integrated into an automation solution. We describe the methodology for developing different models that already exist in elementary process engineering (fluid flow, energy and mass transfer and chemical processes) based on the concepts introduced in Chapter 1. First of all, we will evaluate the often nonlinear models of knowledge obtained from laws of physics; next, we will calculate tangent behavior models, which are represented by differential equations or by transfer functions in order to obtain the control models that are used for designing control algorithms. Sometimes the control model results from an identification operation; although the identified model has an abstract form, it expresses the approximate dynamics of the controlled process. After evaluating this last class of models, we take a look at computing the algorithm for systems that control various parameters of the process, proposing the best methods for designing traditional PID and polynomial algorithms, which are easy to implement in practice. For control, we consider some representative processes of fluid flow, supply drain of liquids, thermoenergy transfer and chemical processes.

7.1. Modeling and control of fluid transfer processes

To design fluid transfer process control systems in a real application, two steps must be followed: evaluate the dynamic models of these types of processes and determine a model-based control algorithm that ensures the desired performances.

7.1.1. *Modeling fluid flow processes*

Fluid flow processes are found in most industrial applications; fluid flow modeling is based on the mechanical action of the forces involved in the process

[TER 98, DAU 04, POP 06, POP 11]. To calculate the model, we must consider a section of a pipe delimited by the actuator and the flow sensor.

We assume that the fluid passing through the pipe is incompressible and the pulse conservation equation is then applied to the system for two different scenarios, which are the most common ones in practice: short pipes with $L \cong D$ and long pipes with $L \gg D$, where L is the length and D is the diameter of the pipe.

7.1.1.1. *Dynamic model of a short pipe*

Let us consider the pipe in Figure 7.1, where Q is the flow rate, ΔP is the differential pressure across the restriction, α is the flow coefficient, S is the flow section and ρ is the fluid density. In this case, the short pipe in question is equivalent to a hydraulic resistance for which the following relation is valid:

$$Q = \alpha\sqrt{\frac{2\Delta P}{\rho}}\,. \qquad [7.1]$$

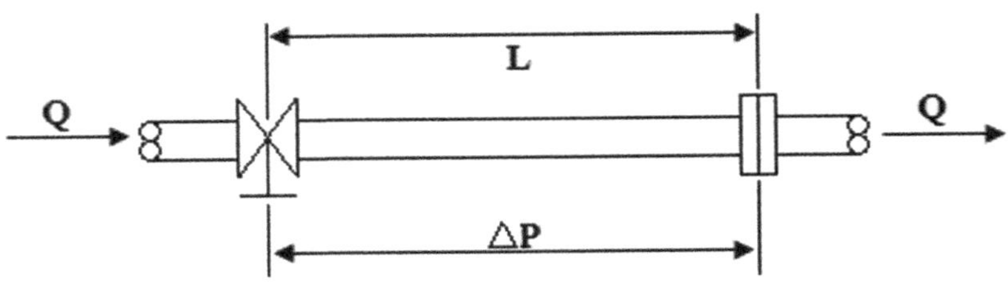

Figure 7.1. *Fluid flow pipe*

For liquid flow in stationary regime, the forces acting on the system are in equilibrium:

$$\Delta P_0 S - \frac{Q_0^2 \rho}{2\alpha^2 S^2} S = 0 \qquad [7.2]$$

where $\Delta P_0 S$ is the active force of the liquid and $\frac{Q_0^2 \rho}{2\alpha^2 S^2} S$ is the reactive force due to the restriction.

In the dynamic state, the difference between both forces is compensated by the temporal variation of the quantity of movement within the system:

$$\Delta P(t)S - \frac{Q^2(t)\rho}{2\alpha^2 S^2}S = \frac{d}{dt}(Mv). \qquad [7.3]$$

In equation [7.3], M represents the mass of the fluid in the pipe and v is its displacement velocity. So we obtain:

$$\Delta P(t)S - \frac{Q^2(t)\rho}{2\alpha^2 S^2}S = \rho LS\frac{1}{S}\frac{dQ(t)}{dt}. \qquad [7.4]$$

In general, the model revolves around an operating point on small variations that occur around this point, which translates to:

$$\begin{aligned} &\Delta P(t) = \Delta P_0 + \Delta(\Delta P(t)) = \Delta P_0 + \Delta p(t) \\ &Q(t) = Q_0 + \Delta Q(t) \end{aligned}. \qquad [7.5]$$

From equations [7.4] and [7.5], we obtain:

$$[\Delta P_0 + \Delta p(t)]S - \frac{[Q_0 + \Delta Q(t)]^2 \rho}{2\alpha^2 S^2}S = \rho L\frac{d[Q_0 + \Delta Q(t)]}{dt}. \qquad [7.6]$$

If we extract the stationary regime expressed by [7.2], we obtain:

$$\Delta p(t)S - \frac{2\rho Q_0 \Delta Q(t) + \rho \Delta Q(t)^2}{2\alpha^2 S^2}S = \rho L\frac{d[\Delta Q(t)]}{dt}. \qquad [7.7]$$

In this equation, which corresponds to the nonlinear model of knowledge, $\Delta p(t)$ represents the variation $\Delta(\Delta P(t))$ of the differential pressure.

We ignore the quadratic term in $\Delta Q^2(t)$ in equation [7.7] and, using the normalized values for $Q(t)$ and $p(t)$ with respect to their values in steady state, we obtain:

$$\begin{aligned} y(t) &= \frac{\Delta Q(t)}{Q_0} \\ u(t) &= \frac{\Delta p(t)}{\Delta P_0} \end{aligned}. \qquad [7.8]$$

The non-dimensional *tangent* model, which expresses the behavior of the flow process, is then the following:

$$\alpha^2 \frac{V_0}{F_0}\frac{dy(t)}{dt} + y(t) = \frac{1}{2}u(t)\,, \qquad [7.9]$$

where V_0 represents the volume of fluid in the pipe in steady state.

From differential equation [7.9], we can deduce the transfer function of the dynamic model:

$$G_{pa}(s) = \frac{k_p}{\tau_{pa}s+1}\,, \qquad [7.10]$$

Where K_p is the gain ($k_p = 0.5$) and τ_{pa} is the time constant of the linearized model $\tau_{pa} = \alpha^2 \frac{V_0}{F_0}$.

The corresponding discrete time model is:

$$y(k+1) - e^{-\frac{h}{\tau_{pa}}} y(k) = k_p\left(1 - e^{-\frac{h}{\tau_{pa}}}\right)u(k) \qquad [7.11]$$

and the resulting discrete transfer function is:

$$G_{pa}(z^{-1}) = \frac{k_p\left(1-e^{-\frac{h}{\tau_{pa}}}\right)z^{-1}}{1-z^{-1}e^{-\frac{h}{\tau_{pa}}}} = \frac{0.5\left(1-e^{-\frac{hF_0}{\alpha^2 V_0}}\right)z^{-1}}{1-z^{-1}e^{-\frac{hF_0}{\alpha^2 V_0}}} = \frac{b_1 z^{-1}}{1+a_1 z^{-1}}\,,$$

where the sampling period h respects the condition $h \le 0{,}1\tau_{pa}$.

For a compressible fluid, we carry out a similar calculation. Nevertheless, equation [7.1] is corrected with a known compressibility coefficient, which will multiply the expression of flow Q [POP 06, POP 11].

7.1.1.2. *Dynamic model of a long pipe*

In this case, it is assumed that the reaction force is the frictional force of the fluid along the length of the pipe, as the flow rate is essentially dependent on length L :

$$Q = L^2 \sqrt{\frac{\Delta P}{k\rho}} \,. \qquad [7.12]$$

For the stationary state of flow, we have an equilibrium of forces within the system:

$$\Delta P_0 S - k\rho LS \frac{Q_0^2}{L^5} = 0 \,, \qquad [7.13]$$

where k is the coefficient of fluid friction in the pipe.

In dynamical regime, we write:

$$\Delta P(t) S - k\rho LS \frac{Q^2(t)}{L^5} = \frac{d(Mv(t))}{dt} \,. \qquad [7.14]$$

The variables $\Delta P(t)$ and $Q(t)$ have the same meanings as in expression [7.2].

Equation [7.3] becomes:

$$[\Delta P_0 + \Delta p(t)]S - k\rho LS \frac{[Q_0 + \Delta Q(t)]^2}{L^5} = \rho LS \frac{1}{S} \frac{d[Q_0 + \Delta Q(t)]}{dt} \,. \qquad [7.15]$$

If we extract the stationary state expressed by [7.13] in equation [7.15], we obtain:

$$\Delta p(t) S - k\rho LS \frac{2Q_0 \Delta Q(t) + \Delta Q(t)^2}{L^5} = \rho L \frac{d[\Delta Q(t)]}{dt} \qquad [7.16]$$

and if we also ignore the quadratic term $\Delta Q^2(t)$, the equation then becomes:

$$\Delta p(t) S - \frac{2k\rho LSQ_0 \Delta Q(t)}{L^5} = \rho L \frac{d[\Delta Q(t)]}{dt} \,. \qquad [7.17]$$

Using normalized variables that are similar to the mechanism mentioned in the previous example, we obtain the following input–output model:

$$\frac{L^5}{2kQ_0S}\frac{dy(t)}{dt}+y(t)=\frac{1}{2}u(t)$$

from which we can deduce the transfer function

$$G_{pb}(s)=\frac{k_p}{\tau_{pb}s+1} \quad [7.18]$$

with $k_p=0.5$ and $\tau_{pb}=\dfrac{L^5}{2kQ_0S}$.

Let us consider a sampling period h (of around a second) to compute the discrete model for [7.18]. We thus get the following discrete transfer function, which represents the process model:

$$G_{pb}(z^{-1})=\frac{b_1z^{-1}}{1+a_1z^{-1}} \quad [7.19]$$

with $b_1=k_p\left(1-e^{-\frac{h}{\tau_{pb}}}\right)$ and $a_1=e^{-\frac{h}{\tau_{pb}}}$.

7.1.2. *Designing flow control systems*

A few observations should be considered when designing a flow control system. This type of system is characterized by a small inertia, so the system is often disturbed – the recommendation is to avoid the derivative component of the control algorithm, which can amplify the effect of the disturbances. More efficient closed-loop systems (for example, cascade configuration of level flow, temperature flow, etc.) have structures that use the flow control loop as a secondary loop; the idea involves ensuring that for this looped secondary system there is a dynamic behavior that gently modifies the dynamics of the main system (the principal loop); the calculation strategy for the flow loop implies specifying faster and aperiodic performances [POP 06].

Taking into account these remarks and the simplified form of the dynamic control model, the usual algorithm for fluid flow control is a traditional PI algorithm.

To design the system, we need the following information: the overall control model that will integrate the dynamics of the actuator (proportional element), the process dynamics (pipe) that are expressed by a first-order model as calculated by equation [7.10] from the flow sensor (proportional element) and also the performances imposed for the closed-loop system.

The dynamic model of the system is then written as:

$$G_f(s) = \frac{k_f}{T_f s + 1}, \quad [7.20]$$

where k_f is the gain of the model (k_f is the product between the actuator, process and sensor gains), T_f is the time constant of the process $\left(T_f = \alpha^2 \frac{V_0}{Q_0}\right)$ and the proposed algorithm is of PI type,

$$G_c = \frac{k_r(T_i s + 1)}{T_i s}. \quad [7.21]$$

The desired (aperiodic) dynamic performance of the closed-loop system is imposed simply by the transfer function:

$$G_{0d} = \frac{1}{T_0 s + 1}, \quad [7.22]$$

where the time constant T_0 ensures the settling time of the dynamic response of the closed-loop system.

The open-loop transfer function is:

$$G(s) = G_c G_p = \frac{k_r(T_i s + 1)}{T_i s} \frac{k_f}{T_f s + 1} \quad [7.23]$$

and the transfer function of the closed loop system is:

$$G_o = \frac{k_r k_f (T_i s + 1)}{T_i s(T_f s + 1) + k_r k_f (T_i s + 1)}. \quad [7.24]$$

The characteristic polynomial of the system is written as:

$$P_c = T_i s(T_f s + 1) + k_r k_f (T_i s + 1) = T_i T_f s^2 + T_i s(k_r k_f + 1) + k_r k_f \,. \qquad [7.25]$$

To ensure aperiodic dynamics expressed by structure [7.22], we impose real negative poles using the roots of the following expression:

$$T^2{}_i s^2 (k_r k_f + 1)^2 - 4 T_i T_f k_r k_f = 0 \,. \qquad [7.26]$$

Having studied the space of controller parameters (k_r, T_i) using the expression $T_i = f\,(k_r)$, in order to obtain the fastest response of the closed loop system, we obtain the following parameter values:

$$k_r^* = \frac{1}{k_f}, T_i^* = \frac{T_f}{k_r k_f} \qquad [7.27]$$

For the discretized form of the PI algorithm, we obtain:

$$\frac{R(z^{-1})}{S(z^{-1})} = \frac{r_0 + r_1 z^{-1}}{1 - z^{-1}}$$

It is then easy to determine the parameters r_0, r_1, as a function of k_r^* and T_i^*, calculated by equations [7.27] [LAN 95].

7.2. Modeling and controlling liquid storage processes

The fluid level dynamics in a tank can be described by modeling a storage process [TER 98, DAU 04, POP 06, POP 11].

Let us consider the dynamic model of this process for a constant section of the tank S supplied with flow rate Q_a and with output flow rate Q_e (see Figure 7.2). Two situations are possible:

– constant output flow at $Q_e = ct$;

– variable output flow, depending on the level of fluid in the reservoir, $Q_e(L)$.

The equations of this model are based on the conservation of the quantity of fluid conveyed throughout the process.

7.2.1. *Constant output flow*

In the stationary regime (zero accumulation in the system), the inlet and outlet flow rates are equal, thus:

$$\rho Q_{a0} - \rho Q_{e0} = 0, \qquad [7.28]$$

where Q_{a0} is the input flow, Q_{e0} is the output flow and ρ is density of the liquid.

In the dynamical state, the variation of the mass in the system is equal to the difference of the incoming and outgoing flows:

$$\rho Q_a(t) - \rho Q_{e0} = \frac{dM(t)}{dt} = \rho S \frac{dL(t)}{dt}. \qquad [7.29]$$

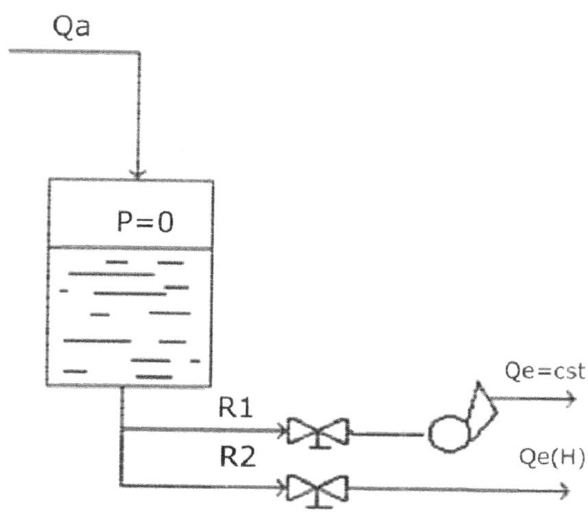

Figure 7.2. *Representation of a storage process*

In equation [7.29], S represents the constant section of the tank, $L(t)$ is the level of the liquid at time t and $M(t)$ is the mass of the liquid in the reservoir at time t. We assume that the system will evolve around an operating point that corresponds to a steady state:

$$L(t) = L_0 + \Delta L(t)$$
$$F_a(t) = F_{a0} + \Delta F_a(t) \qquad [7.30]$$

From [7.29] and [7.30], we obtain:

$$\rho[Q_{a0} - \Delta Q_a(t)] - \rho Q_{e0} = \frac{dM(t)}{dt} = \rho S \frac{d[L_0 + \Delta L(t)]}{dt}. \qquad [7.31]$$

And by taking into account the condition of the stationary state expressed by [7.28], we obtain:

$$\Delta Q_a(t) = S \frac{d\Delta L(t)}{dt} \qquad [7.32]$$

By normalizing around steady-state values, we obtain the resulting dimensionless variables:

$$\frac{\Delta L(t)}{L_0} = y(t)$$
$$\frac{\Delta Q_a(t)}{Q_{a0}} = u(t).$$

Thus, equation [7.32] becomes the tangent behavior model:

$$\frac{SL_0}{Q_{a0}} \frac{dy(t)}{dt} = \frac{V_0}{Q_{a0}} \frac{dy(t)}{dt} = u(t) \qquad [7.33]$$

or by integration,

$$y(t) = \frac{Q_{a0}}{V_0} \int_0^t u(t)dt. \qquad [7.34]$$

The input–output transfer function can easily be deduced:

$$H_{pa}(s) = \frac{1}{\tau_{pa} s}, \qquad [7.35]$$

with $\tau_{pa} = \dfrac{V_0}{Q_{a0}}$.

The discretization of expression [7.35] leads to a discrete linear model, which is expressed by the transfer function:

$$H_{pa}(z^{-1}) = \frac{h}{\tau_{pa}(1 - z^{-1})}.$$

where h is the chosen sampling period (in practice, around tens of seconds).

Equations [7.34] and [7.35] show that the constant output flow process behaves like an integrator, which is why it is called a "*process without self-stabilization*".

7.2.2. *Variable output flow*

For a stationary process regime, equation [7.28] remains valid. In a dynamical regime, writing out the mass balance gives the following equation:

$$\rho Q_a(t) - \rho Q_e(t) = \frac{dM(t)}{dt} = \rho S \frac{dL(t)}{dt}. \qquad [7.36]$$

Here, flow rate Q_e depends on level L in the tank, according to a relation of the type:

$$Q_e = \alpha\sqrt{2gL}, \qquad [7.37]$$

where α is the output flow section and g is the gravitational acceleration. Its development in a Taylor series around an operating point is written as:

$$Q_e = Q_{e0} + \left(\frac{\partial Q_e}{\partial L_0}\right)\frac{L - L_0}{1!} + \left(\frac{\partial^2 Q_e}{\partial L_0^2}\right)\frac{(L - L_0)^2}{2!} + \ldots. \qquad [7.38]$$

For the linear part, we obtain the equation:

$$Q_e \cong Q_{e0} + \left(\frac{\partial Q_e}{\partial L_0}\right)(L - L_0) \qquad [7.39]$$

which can be written as:

$$Q_e(t) - Q_{e0} = \left(\frac{\partial Q_e}{\partial L_0}\right)\left(L(t) - L_0\right) \qquad [7.40]$$

or even

$$\Delta Q_e(t) = \left(\frac{\partial Q_e}{\partial L_0}\right)\Delta L(t). \qquad [7.41]$$

Let us again consider equation [7.33] and the time-varying variables, as expressed by:

$$\begin{aligned} L(t) &= L_0 + \Delta L(t) \\ Q_a(t) &= Q_{a0} + \Delta Q_a(t). \\ Q_e(t) &= Q_{e0} + \Delta Q_e(t) \end{aligned}$$

We then get:

$$[Q_{a0} + \Delta Q_a(t)] - [Q_{e0} + \Delta Q_e(t)] = S\frac{d[L_0 + \Delta L(t)]}{dt}. \qquad [7.42]$$

Considering the steady-state equation [7.28], we obtain the relation:

$$\Delta Q_a(t) - \Delta Q_e(t) = S\frac{d[\Delta L(t)]}{dt}. \qquad [7.43]$$

On the other hand, we obtain the dimensionless variables of the model:

$$\begin{aligned} \frac{\Delta L(t)}{L_0} &= y(t) \\ \frac{\Delta Q_a(t)}{Q_{a0}} &= u(t) \end{aligned}.$$

If we consider equation [7.41], by replacing the derivative of Q_e with respect to L, and after the usual calculation in [7.43], we obtain the final formula for this equation:

$$\frac{2V_0}{Q_{a0}}\frac{dy(t)}{dt} + y(t) = 2u(t). \qquad [7.44]$$

By means of the above simplifications, we obtain a simple transfer function for the elementary storage process:

$$H_{pb}(s) = \frac{k_{pb}}{\tau_{pb}s+1}, \quad [7.45]$$

where $k_{pb} = 2$ and $\tau_{pb} = 2\frac{V_0}{Q_{a0}}$ are the parameters of the model.

The discrete first-order equivalent model is written as:

$$H_{pb}(z^{-1}) = \frac{b_1 z^{-1}}{1+a_1 z^{-1}} = \frac{k_{pb}(1-e^{-h/\tau_{pb}})z^{-1}}{1-e^{-h/\tau_{pb}}z^{-1}}.$$

We can say that close to a stationary point, the storage process with variable output flow (with respect to level L) behaves like a first-order system, which is why it called a "*process with self-stabilization*".

7.2.3. *Designing liquid level control systems*

To design a closed-loop system that controls the process with constant output flow, we have at our disposal the overall control model. This model includes the actuator dynamics (the proportional element), the process dynamics (expressed by the integrator calculated in equation [7.35]) and the level sensor dynamics (proportional element).

The overall model is expressed by the following equation:

$$G_f(s) = k_a \frac{1}{\tau_{pa}s} k_t = \frac{k_f}{s} \quad [7.46]$$

where k_f is the multiplication of the gains from the actuator, the model of the process and the sensor. Thus, this transfer function is an integrator and the performance of the system in the closed-loop system is expressed by an imposed dynamic:

$$G_{0e}(s) = \frac{1}{T_0 s+1}. \quad [7.47]$$

It is easy to see that the main objective is to put forward a stabilizing control for the closed-loop system, and that the proportional control algorithm can solve this objective for the model [7.46].

Let us consider a proportional controller K_r, for which we have:

$$G_0(s) = \frac{k_r k_f}{s + k_r k_f}. \quad [7.48]$$

The proportional control algorithm that ensures the imposed performance is then

$$k_r^* = \frac{T_0}{k_f}.$$

For a process with variable (free) output flow, a closed-loop system that controls the storage process is designed in the same way. The information necessary to evaluate the control law is then: the overall control model, which includes the dynamics of the actuator (first-order element); the process dynamics, which is also expressed by a first-order system calculated by equation [7.45]; the gain of the level sensor (proportional element) and the dynamic performances of the closed-loop system.

This time, in the mentioned hypotheses, the overall model of the process is expressed by the second-order transfer function:

$$G_p = \frac{k_{pa}}{\tau_{pa}s+1}\frac{k_{pb}}{\tau_{pb}s+1}k_t = \frac{k_f}{\left(\tau_{pa}s+1\right)\left(\tau_{pb}s+1\right)}. \quad [7.49]$$

We consider that the standard representation of a second-order closed-loop dynamic system, defined by:

$$G(s) = \frac{\omega_n^2}{s^2 + 2\zeta\omega_n s + \omega_n^2} \quad [7.50]$$

will ensure the desired performance through the selected imposed pair (ζ, ω_n).

We can calculate the second-order discrete polynomial that corresponds to the denominator of the transfer function [7.50]:

$$P(z^{-1}) = 1 + p_1 z^{-1} + p_2 z^{-2}. \quad [7.51]$$

On the other hand, the discrete equivalent of the continuous model [7.49] of the process is:

$$\frac{\hat{b}(z^{-1})}{\hat{a}(z^{-1})}.$$

Let us consider a PI controller in discrete representation:

$$\frac{R(z^{-1})}{S(z^{-1})} = \frac{r_0 + r_1 z^{-1}}{1 - z^{-1}}. \qquad [7.52]$$

The polynomial equation that gives the unknown parameters of the controller is given as:

$$\hat{a}(z^{-1})S(z^{-1}) + \hat{b}(z^{-1})R(z^{-1}) = 1 + p_1 z^{-1} + p_2 z^{-2}.$$

The solution of this polynomial equation represents the discretized controller

$R(z^{-1}), S(z^{-1})$, that ensures the desired performances:

$$\begin{aligned} R(z^{-1}) &= r_0 + r_1 z^{-1} = (p_1 - a_1 - 1)/b_1 + [(p_2 + a_1)/b_1]z^{-1} \\ S(z^{-1}) &= 1 - z^{-1} \end{aligned}.$$

7.3. Modeling and controlling the storage process of a pneumatic capacitor

Pressure is an important parameter for characterizing certain pneumatic or hydraulic systems, but it is also important for energy and chemical installations.

7.3.1. *Modeling a pneumatic capacitor*

In what follows, we define the mathematical model of a pneumatic capacitor with constant volume V, in which there is a storage gas at temperature T and pressure p, with input gas flow Q_a and output gas flow Q_e (Figure 7.3). If we assume that the gas acts like a perfect gas and that temperature is constant, we can then apply the ideal gases law:

$$pV = MRT, \qquad [7.53]$$

where M is the mass of gas in volume V and R is the ideal gas constant.

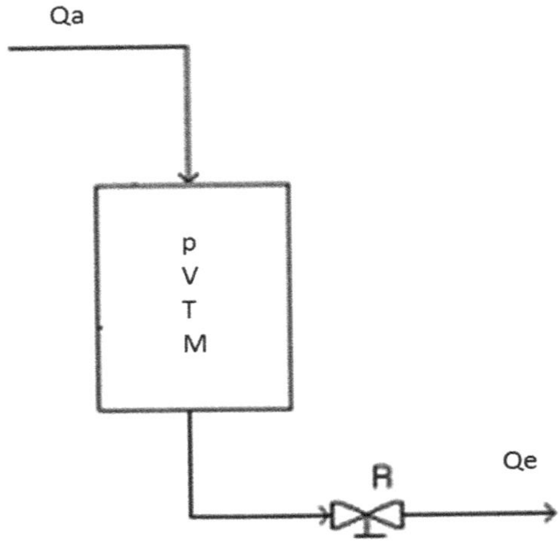

Figure 7.3. *Representation of a pneumatic process*

Under these assumptions, the pressure variation of the capacitor depends on the mass variation. Indeed, the derivation with respect to time in equation [7.53] gives:

$$V\frac{dp(t)}{dt} = RT\frac{dM(t)}{dt} \qquad [7.54]$$

but,

$$\frac{dM(t)}{dt} = Q_a(t) - Q_e(t), \qquad [7.55]$$

so,

$$\frac{dp(t)}{dt} = \frac{RT}{V}\left[Q_a(t) - Q_e(t)\right]. \qquad [7.56]$$

Let us assume that the output flow depends on pressure according to the equation (laminar flow): $Q_e = \alpha\sqrt{p(p-p_c)}$ where a is a constant determined by the pneumatic resistance of the output path and p_c is the pressure on the consumer.

Using the Taylor series development expansion around the nominal operating point, we find:

$$Q_e = Q_{e0} + \left(\frac{\partial Q_e}{\partial p_0}\right)\frac{p-p_0}{1!} + \left(\frac{\partial^2 Q_e}{\partial p_0^2}\right)\frac{(p-p_0)^2}{2!} + \ldots. \quad [7.57]$$

The linear part of equation [7.57] for pressure p is written as:

$$Q_e = Q_{e0} + \left(\frac{\partial Q_e}{\partial p_0}\right)(p-p_0) \quad [7.58]$$

or even,

$$\Delta Q_e(t) = \left(\frac{\partial Q_e}{\partial p_0}\right)\Delta p(t). \quad [7.59]$$

Let us consider the variables over time, as expressed by:

$$\begin{aligned} p(t) &= p_0 + \Delta p(t) \\ Q_a(t) &= Q_{a0} + \Delta Q_a(t). \\ Q_e(t) &= Q_{e0} + \Delta Q_e(t) \end{aligned} \quad [7.60]$$

Considering [7.59], [7.60] and the steady-state condition, we obtain:

$$\Delta Q_a(t) - \left(\frac{\partial Q_e}{\partial p_0}\right)\Delta p(t) = \frac{V}{RT}\frac{d[\Delta L(t)]}{dt}. \quad [7.61]$$

Through the same normalization procedure, we get the dimensionless variables of the model:

$$\begin{aligned} \frac{\Delta p(t)}{p_0} &= y(t) \\ \frac{\Delta Q_a(t)}{Q_{a0}} &= u(t) \end{aligned}.$$

The final dynamic model is:

$$\frac{V}{RT}\left(\frac{\partial Q_e}{\partial p_0}\right)^{-1}\frac{dy(t)}{dt} + y(t) = \frac{\partial Q_a}{\partial p_0}\left(\frac{\partial Q_e}{\partial p_0}\right)^{-2}u(t). \quad [7.62]$$

Thus, the transfer function of the continuous model is:

$$H_p(s) = \frac{k_p}{\tau_p s + 1} \qquad [7.63]$$

and the corresponding discrete transfer function is:

$$H_p(z^{-1}) = \frac{b_1}{1 + a_1 z^{-1}} = \frac{k_p \left(1 - e^{-\frac{h}{\tau p}}\right) z^{-1}}{1 - e^{-\frac{h}{\tau p}} z^{-1}} \qquad [7.64]$$

where,

$$k_p = \frac{Q_a}{p_0} \left(\frac{\partial Q_e}{\partial p}\right)_0^{-1}, \tau_p = \frac{V}{RT} \left(\frac{\partial Q_e}{\partial p}\right)_0^{-1} \qquad [7.65]$$

are the parameters of the behavior model and h is the sampling period (around a second).

7.3.2. *Designing pneumatic capacitor control systems*

The information needed to be able to design the control algorithm for this type of process is the overall model that includes the dynamics of the actuator assembly (proportional), the pneumatic capacitor, the pressure sensor (proportional) and the performance for the closed-loop system.

The transfer function of the overall control model is written as:

$$G(s) = k_a \frac{k_p}{\tau_p s + 1} k_t = \frac{k_f}{\tau_f s + 1} \qquad [7.66]$$

Based on the same motivation as in flow control, a PI algorithm is recommended to ensure the performance of the closed-loop system. The control approach used here is optimal and represents the solution for minimizing an integral criterion, which is expressed as a function of the regulation error (epsilon(t)) and its derivative:

$$\min \left\{ J = \int_0^\infty [\varepsilon(t) + q\dot{\varepsilon}(t)] dt \right\} \qquad [7.67]$$

For model [7.66] and the PI controller, the direct expression of optimal criterion [7.67], which is calculated with the parameters of the controller and model, becomes a nonlinear programming problem.

$$\min\left\{J=\frac{T_i+K_rKT}{2K_rK(1+K_rK)}+q\frac{K_rK\left(1+K_rK\dfrac{T_i}{T}\right)}{2T_i(1+K_rK)}\right\}, \qquad [7.68]$$

where q is a criterion-weighting coefficient.

This problem is solved for the predetermined model (known k_f and τ_f) through a digital gradient technique in the parameter space of the controller:

$$\min\{J(k_f,\tau_f,K_r,T_i)\}.$$

The solution to this problem gives the optimal parameters of the control algorithm (K_r^*, T_i^*).

This approach makes it possible to introduce the concept of duality, meaning to re-identify the model in real time if necessary (if the process changes its operating point), using the criterion [7.68] according to parameters (k_f, τ_f), or to recalculate the control algorithm (values K_r^*, T_i^*) in the parameter space of the controller for the well-defined model (known approach – partition of variables of the criterion function). The new model recursively imposes a new control algorithm to preserve the performance [CAL 79, DAU 04, POP 06, BOR 13].

7.4. Modeling and controlling heat transfer processes

Temperature is a representative variable for industrial processes of heat transfer [TER 98, DAU 04, POP 06, POP 11]. In this case, we are interested in modeling heat transfer from a thermal agent to a product to be heated or cooled.

7.4.1. *Modeling a thermal transfer process*

We seek to develop a mathematical model for a heat transfer process by convection of the thermal agent and the liquid phase product (Figure 7.4) in a volume V characterized by the following variables: Q_{a0} is the thermal agent flow rate, T_a is the temperature of the thermal agent, c_a is the specific heat of the thermal agent, Q_p is the product flow, T_p is the product temperature; c_p is the specific heat of the product, ρ_a is the density of the thermal agent, ρ_p is the density of the product, T_e is the temperature of the mixture, C_e is the specific heat of the mixture and ρ_e is the density of the mixture.

In the stationary regime, the energy balance equation (ignoring external heat loss) is written as:

$$\rho_a Q_{a0} c_a T_a + \rho_p Q_p c_p T_p - \rho_e Q_{e0} c_e T_{e0} = 0 \qquad [7.69]$$

and the additional equation for the conservation of mass is:

$$Q_{e0} = Q_{a0} + Q_p .$$

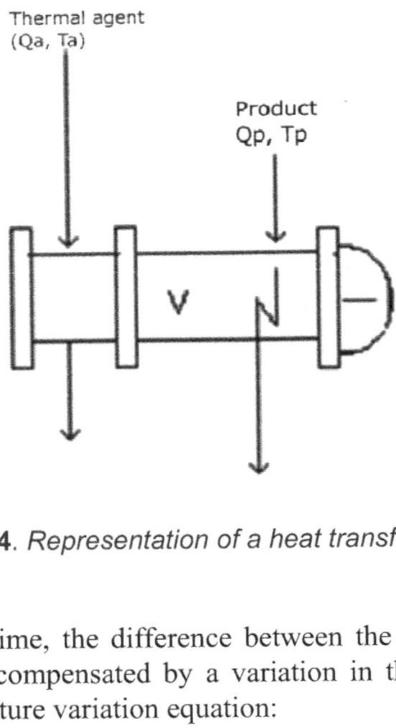

Figure 7.4. *Representation of a heat transfer process*

In the dynamical regime, the difference between the input and output thermal fluxes of the system is compensated by a variation in the amount of heat that is expressed by the temperature variation equation:

$$\rho_a Q_a(t) c_a T_a + \rho_p Q_p c_p T_p - \rho_e Q_e(t) c_e T_e(t) = \rho_e V c_e \frac{dT_e(t)}{dt} \qquad [7.70]$$

$$Q_e(t) = Q_a(t) + Q_p .$$

Parameters that are variable over time are written as:

$$T_e(t) = T_{e0} + \Delta T_e(t)$$
$$Q_a(t) = Q_{a0} + \Delta Q_a(t)$$

giving the following equation:

$$\begin{aligned}&\rho_a[Q_{a0} + \Delta Q_a(t)]c_a T_a + \rho_p Q_p c_p T_p \\ &-\rho_e[Q_{a0} + \Delta Q_a(t) + Q_p]c_e[T_{e0} + \Delta T_e(t)] = \rho_e V c_e \frac{d[T_{e0} + \Delta T_e(t)]}{dt} \cdot\end{aligned} \quad [7.71]$$

By eliminating the stationary state expressed by equations [7.69] and ignoring the higher order terms, we obtain:

$$\begin{aligned}&\rho_a \Delta Q_a(t) c_a T_a - \rho_e Q_{a0} c_e \Delta T_e(t) - \rho_e T_{e0} c_e \Delta Q_a(t) \\ &= \rho_e V c_e \frac{d[\Delta T_e(t)]}{dt}\end{aligned} \quad [7.72]$$

where

$$\rho_e V c_e \frac{d[\Delta T_e(t)]}{dt} + \rho_e Q_{a0} c_e \Delta T_e(t) = (\rho_a c_a T_a - \rho_e T_{e0} c_e)\Delta Q_a(t). \quad [7.73]$$

Through the same normalization technique as before, we obtain the dimensionless variables of the model:

$$\begin{aligned}&\frac{\Delta T_e(t)}{T_{e0}} = y(t) \\ &\frac{\Delta Q_a(t)}{Q_{a0}} = u(t)\end{aligned} \quad . \quad [7.74]$$

If we introduce [7.74] into [7.73], we obtain, after a very simple calculation:

$$\frac{V}{Q_{e0}} \frac{dy(t)}{dt} + y(t) = \frac{\rho_a c_a T_a - \rho_e c_e T_{e0}}{\rho_e c_e T_{e0}} \frac{Q_{a0}}{Q_{e0}} u(t). \quad [7.75]$$

The transfer function that corresponds to this differential equation is written as:

$$H_p(s) = \frac{k_p}{\tau_p s + 1} \quad [7.76]$$

with the behavior model parameters:

$$k_p = \frac{\rho_a c_a T_a - \rho_e c_e T_{e0}}{\rho_e c_e T_{e0}} \frac{Q_{a0}}{Q_{e0}}, \qquad \tau_p = \frac{V}{Q_{e0}}.$$

7.4.2. *Designing temperature control systems*

Industrial processes with heat transfer are often very slow. We consider that the actuator is approximated by a proportional element and the temperature sensor has the dynamics of a first-order system.

Under these conditions, the transfer function of the process model is:

$$G_p(s) = k_a \frac{k_p}{\left(\tau_p s + 1\right)} \frac{k_t}{\left(T_t s + 1\right)} = \frac{k_f}{(\tau_p s + 1)(T_t s + 1)}. \qquad [7.77]$$

We recommend using a proportional-integral-derivative PID algorithm to design the control system. We work directly in discrete form by using the pole placement method. The performances of the closed-loop system are imposed by a second-order characteristic polynomial $P_i(z^{-1})$, which is specified in advance by the two desired poles (coefficients p_1, p_2 are therefore known):

$$P_i(z^{-1}) = 1 + p_1 z^{-1} + p_2 z^{-2}. \qquad [7.78]$$

The discretized form of equation [7.77], obtained using a sampling period of a few tens of seconds, is written as:

$$G_p(z^{-1}) = \frac{\hat{b}(z^{-1})}{\hat{a}(z^{-1})} \qquad [7.79]$$

The PID algorithm with filtering on the derivative component is described by the transfer function:

$$G_c(z^{-1}) = \frac{R(z^{-1})}{S(z^{-1})} = \frac{r_0 + r_1 z^{-1} + r_2 z^{-2}}{(1 - z^{-1})(1 + s_1 z^{-1})}. \qquad [7.80]$$

The transfer function of the closed-loop system is thus:

$$G_0(z^{-1}) = \frac{\hat{b}(z^{-1}) R(z^{-1})}{\hat{a}(z^{-1}) S(z^{-1}) + b(z^{-1}) R(z^{-1})} \qquad [7.81]$$

and the characteristic polynomial of the system is therefore:

$$P(z^{-1}) = \hat{a}(z^{-1})S(z^{-1}) + \hat{b}(z^{-1})R(z^{-1}). \qquad [7.82]$$

We impose the polynomial identity $P(z^{-1}) = P_i(z^{-1})$ and these results in the Diophantine equation:

$$\hat{a}(z^{-1})S(z^{-1}) + \hat{b}(z^{-1})R(z^{-1}) = 1 + p_1 z^{-1} + p_2 z^{-2} \qquad [7.83]$$

with $R(z^{-1})$ and $S(z^{-1})$ being the unknown polynomials of the PID controller.

The control algorithm that ensures the specified performance is given by the solution to equation [7.83] represented by the two polynomials $R^*(z^{-1}), S^*(z^{-1})$, [FOU 04, POP 06].

7.5. Modeling and control of component transfer processes

Concentration is a parameter that is linked to chemical processes with or without reaction. These usually occur in chemical or petrochemical facilities. There is a wide variety of reactors of various shapes and/or sizes. The equation of such processes is based on conservation equations of the mass of components [TER 98, FOU 04, POP 06, POP 11].

The balance equations written above take different forms depending on the system being considered and the mode of operation. In this case, we mean any reactor where there is a transfer of material with or without reaction. The balance equations below are written for the two above-mentioned types of chemical processes.

7.5.1. *Modeling a chemical mixing process without reaction*

Here, we are interested in obtaining a substance of intermediate concentration from the mixture of a concentrated substance and another diluted substance in a reactor of constant volume V (Figure 7.5). In steady-state regime of the process being considered, we can write:

$$\rho Q_{d0} c_d + \rho Q_c c_c - \rho Q_{e0} c_{e0} = 0 \qquad [7.84]$$

$$Q_{d0} + Q_{c0} = Q_{e0}, \qquad [7.85]$$

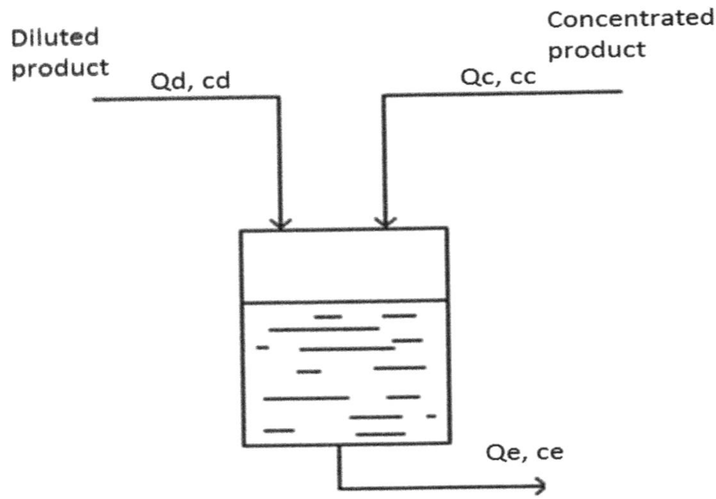

Figure 7.5. *Representation of a process without reaction*

where Q_{d0} is the flow rate of diluent, Q_c is the flow rate of the concentrated substance, c_d is the concentration of the diluent, Q_{e0} is the output flow of the mixer, c_{e0} is the concentration of the mixture and ρ is the density of the substance.

The dynamic mixture process in question is described by the equation:

$$\rho Q_d(t)c_d + \rho Q_c c_c - \rho Q_e(t)c_e(t) = \frac{\rho V dc_e(t)}{dt}. \qquad [7.86]$$

Moreover,

$$\begin{aligned} c_e(t) &= c_{e0} + \Delta c_e(t) \\ Q_d(t) &= Q_{d0} + \Delta Q_d(t) \end{aligned}. \qquad [7.87]$$

We substitute [7.86] into [7.85] to obtain:

$$\begin{aligned} &\rho[Q_{d0} + \Delta Q_d(t)]c_d + \rho Q_c c_c - \rho[Q_{e0} + \Delta Q_e(t)][c_{e0} + \Delta c_e(t)] \\ &= \frac{\rho V d[c_{e0} + \Delta c_e(t)]}{dt} \end{aligned} \qquad [7.88]$$

By eliminating the stationary state expressed by equations [7.84] and [7.85] and by ignoring the infinitely small $\Delta Q_d(t)\Delta c_e(t)$, we obtain:

$$c_d \Delta Q_d(t) + Q_c c_c - Q_{e0}\Delta c_e(t) - c_{e0}\Delta Q_d(t) = \frac{V d[\Delta c_e(t)]}{dt}. \qquad [7.89]$$

The dimensionless variables of the model are:

$$\begin{aligned} \frac{\Delta c_e(t)}{c_{e0}} &= y(t) \\ \frac{\Delta Q_d(t)}{Q_{d0}} &= u(t) \end{aligned} . \qquad [7.90]$$

By introducing [7.83] into [7.82], we obtain:

$$\frac{V}{Q_{e0}} \frac{dy(t)}{dt} + y(t) = \frac{c_d - c_{e0}}{c_{e0}} \frac{Q_{d0}}{Q_{e0}} u(t). \qquad [7.91]$$

The transfer function that corresponds to this differential equation is:

$$H_{pa}(s) = \frac{k_{pa}}{\tau_{pa} s + 1}, \qquad [7.92]$$

where,

$$k_{pa} = \frac{c_d - c_{e0}}{c_{e0}} \frac{Q_{d0}}{Q_{e0}} \text{ and } \tau_{pa} = \frac{V}{Q_{e0}}.$$

As a result, this type of process behaves like a first-order system. The corresponding discrete model is presented in the following equation:

$$H_{pa}(z^{-1}) = \frac{b_1 z^{-1}}{1 + a_1 z^{-1}} = \frac{k_{pa}\left(1 - e^{-\frac{h}{\tau a}}\right) z^{-1}}{1 - e^{-\frac{h}{\tau_{pa}}} z^{-1}}$$

with $b_1 = k_{pa}\left(1 - e^{-\frac{h}{\tau_{pa}}}\right)$ and $a_1 = e^{-\frac{h}{\tau_{pa}}}$ (the sampling period h is just a few seconds).

7.5.2. *Modeling a chemical reaction process*

Let us consider a non-isothermal chemical reaction of the type $A \xrightarrow{k} B(A)$, which takes place in a reactor of volume V (substance A partially transforms into B with the reaction rate coefficient k). The mathematical model is determined by the mass conservation equations for reactant A and for reaction product B. The quantities of substances consumed and produced by the reaction are the mass output or input flows into the system (Figure 7.6).

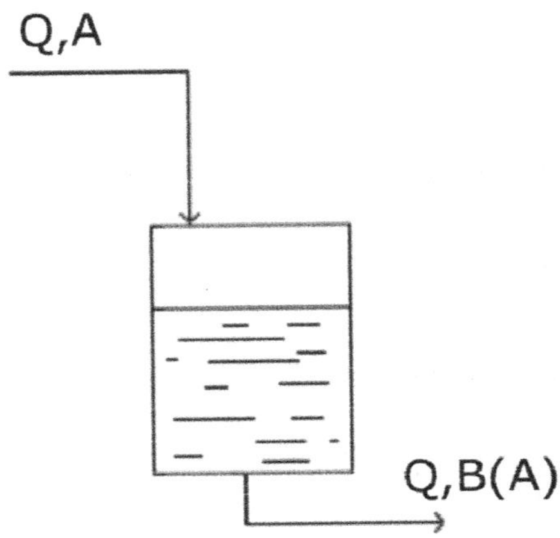

Figure 7.6. *Transfer process of components with chemical reaction*

In the steady-state regime, we then get:

$$Q_0 c_{A1} - kVc_{A2} - Q_0 c_{A2} = 0 \qquad [7.93]$$

$$kVc_{A2} - F_0 c_{B0} = 0. \qquad [7.94]$$

We are interested here in the variation of concentration of the residual substance c_{A2} in an infinitely mixed reactor as a function of variations of inlet and outlet flow rates:

$$Q(t)c_{A1} - kVc_{A2}(t) - Q(t)c_{A2}(t) = \frac{Vdc_{A2}(t)}{dt}. \qquad [7.95]$$

The time-dependent quantities are expressed by the following equations:

$$\begin{aligned} c_{A2}(t) &= c_{A20} + \Delta c_{A2}(t) \\ Q(t) &= Q_0 + \Delta Q(t) \end{aligned} \quad . \tag{7.96}$$

Eliminating the stationary regime and the infinitely small $\Delta F(t)\Delta c_{A2}(t)$ from equations [7.88] and [7.89] allows us to write:

$$c_{A1}\Delta Q(t) - kV\Delta c_{A2}(t) - Q_0\Delta c_{A2}(t) - c_{A20}\Delta Q(t) = \frac{Vd[\Delta c_{A2}(t)]}{dt} \tag{7.97}$$

or even,

$$\Delta Q(t)(c_{A1} - c_{A20}) - (Q_0 + kV)\Delta c_{A2}(t) = \frac{Vd[\Delta c_{A2}(t)]}{dt} \, . \tag{7.98}$$

By normalizing it, we obtain:

$$\begin{aligned} \frac{\Delta c_{A2}(t)}{c_{A20}} &= y(t) \\ \frac{\Delta Q(t)}{Q_0} &= u(t) \end{aligned} \quad . \tag{7.99}$$

With notations [7.99], we obtain the final dimensionless model:

$$Vc_{A20}\frac{dy(t)}{dt} + (Q_0 + kV)c_{A20}y(t) = Q_0(c_{A1} - c_{A20})u(t) \, . \tag{7.100}$$

Then, we obtain:

$$Vc_{A20}\frac{dy(t)}{dt} + y(t) = \frac{Q_0}{(Q_0 + kV)}\frac{(c_{A1} - c_{A20})}{c_{A20}}u(t) \, . \tag{7.101}$$

The following transfer function is made to correspond to the differential equation:

$$H_p = \frac{k_p}{\tau_p s + 1} \, , \tag{7.102}$$

where,

$$k_p = \frac{Q_0}{(Q_0 + kV)} \frac{(c_{A1} - c_{A20})}{c_{A20}}, \quad \tau_p = \frac{V}{(Q_0 + kV)}$$

with the corresponding discrete form:

$$H_p(z^{-1}) = \frac{k_p(1 - e^{-h/\tau_p})z^{-1}}{1 - e^{-h/\tau_p} z^{-1}}, \qquad [7.103]$$

where h is the sampling period (of a few seconds).

7.5.3. *Designing systems for controlling the concentration of chemical components*

In chemical processes, problems exist concerning the acquisition of parameters to be controlled. The sensors are well protected to avoid direct contact with chemical substances through an insulation mechanism that will introduce a pure delay. Thus, the global transfer function of the model is expressed by the following equation:

$$G_p(s) = k_a \frac{k_p}{\tau_p s + 1} k_t (e^{-\tau s}) = \frac{k_f e^{-\tau s}}{\tau_p s + 1}. \qquad [7.104]$$

We propose a PID algorithm and a frequency method for designing the control algorithm. Let us consider the factorized form of the PID algorithm:

$$G_c(s) = \left(\frac{K_r}{T_i s} \right) (T_i s + 1)(T_d s + 1). \qquad [7.105]$$

One can select $T_i^* = \tau_p$ and calculate the transfer function for the open-loop system:

$$G(s) = \frac{K_r}{\tau_p s} (T_d s + 1) k_f e^{-\tau s} = \frac{K}{\tau_p s} (T_d s + 1) e^{-\tau s}. \qquad [7.106]$$

The module and the argument of $G(j\omega)$ are expressed by equations:

$$|G(j\omega)| = \frac{K\sqrt{(T_d \omega)^2 + 1}}{\tau_p \omega} \qquad [7.107]$$

$$\arg G(j\omega) = -\tau\omega + \operatorname{arctg}\left(T_d\omega\right) - \frac{\pi}{2}. \qquad [7.108]$$

The performance of the system is imposed by the modulus and phase margins, which are measured at the Nyquist locus of the open-loop system.

Let us consider a modulus margin of 0.5 and a phase margin of $\frac{\pi}{4}$ at the critical frequency of the system, ω_{cr}.

For the frequency ω_{cr}, equation [7.108] becomes:

$$-\tau\omega_{cr} - \frac{\pi}{2} + \frac{\pi}{4} = -\pi \qquad [7.109]$$

and we calculate the critical frequency ω_{cr},

$$\omega_{cr} = \frac{3\pi}{4\tau}. \qquad [7.110]$$

The derivative constant T_d is easily deduced as a function of the critical frequency from equation [7.108]; gain K_r from the PID controller is calculated using the expression of module [7.107].

$$\begin{aligned} T_d^* &= \frac{4\tau}{3\pi} \\ K_r^* &= \frac{3\sqrt{2}\pi}{16}\frac{\tau_p}{\tau} \end{aligned}. \qquad [7.111]$$

The digital PID control algorithm provided by filtering the derivative component with a sampling period h (of around tens of seconds) uses the result of the design in the continuous state, which is:

$$G_c(z^{-1}) = \frac{R(z^{-1})}{S(z^{-1})} = \frac{r_0 + r_1 z^{-1} + r_2 z^{-2}}{(1 - z^{-1})(1 + s_1 z^{-1})}, \qquad [7.112]$$

where parameters r_0, r_1, r_2 and s_1 are calculated as a function of K_r^*, T_d^*, T_i^* [CAL 79, TER 98, LAN 95]. To go from the factorized form [7.105] to the standard form of the algorithm, a correction is applied to the parameters [POP 06].

8

Industrial Applications – Case Studies

In this chapter, we present the results of a few case studies on how automatic control resources can significantly improve industrial processes exploitation: in the iron and steel industry – an air heating system for supplying the blast furnace of a steel plant; in petrochemical processes – the pyrolysis reactor for making ethylene; in the energy industry – thermal transfer facilities and renewable energy facilities (photovoltaics). It is known that a simple and low-cost automation solution is based on controlling key parameters to ensure a nominal operating point of the process. For implementation, we recommend improved control algorithms using advanced adaptive, robust or predictive strategies to maintain nominal performances in a real industrial setting. On the other hand, we propose corresponding optimization solutions to exploit technological processes in order to increase the profit of the company. The result of the optimization problem, which represents the best decision for the process, is automatically transferred at the control level in order for the process to occur at its optimum operating point. In this context, for each case study, we propose a set of control and decision algorithms that provide independent tracking and regulation performances as well as digital solutions for real-time control and optimization. All the studies presented here have been validated in functioning technological applications.

8.1. Digital control for an installation of air heating in a steel plant

The complexity of metallurgic and steel facilities and their organizational and operational difficulties are well known. Significant improvements in the operation of these facilities were achieved after the emergence of digital control and high-performance computing equipment.

Economically and commercially, the most important evaluation criteria in a steel plant are price, quantity and quality of production. The current priorities in the iron and steel industry are energy saving, productivity and product diversification [WAI 62, TAL 93, POP 06].

Let us consider an air heating system for supplying blast furnaces (cowper installation). The operating mechanism of such an installation comprises three important steps: the actual heating of the air for the blast furnace, the supply of air and the cooling of the installation. By using efficient technological switching through suitable means, the furnace is supplied with hot air (1,200°C). There are some peculiarities for this type of installation (cowpers): the rather large dimensions lead to mathematical models with significant delay or distributed parameters; the fuel used has several components that can sometimes be mixed in a random manner: methane gas, coke gas and blast furnace gas, which all have different heating capacities, as well as nonlinearities of the technological setup. To assess the quality of the combustion, the composition of the residual gas in concentrations of O_2 and CO is measured and analyzed. All these particularities present significant difficulties in the operation of the heating mechanism.

The automatic control configuration proposed for this air heating installation is organized across two interconnected levels in a hierarchical structure: the execution level of the process for parameter control and the supervisory level for optimizing the combustion procedure necessary for heating the air. Let us note that dynamic models of the process control parameters are obtained by experimental recursive identification techniques (recursive least squares (RLS)) and the control algorithms by R (regulation), S (sensitivity) and T (tracking) (RST) polynomials; the decision model is also evaluated by least squares (LS) identification techniques and the optimal decision is calculated by solving a nonlinear programming problem.

8.1.1. *Automation solution and design of the control algorithms*

The main objective is to propose a digital control solution for this type of installation – one that ensures efficient air heating operation for the blast furnace and another that drives the combustion process, which is an important supplier of the required energy. The solution chosen for the control level is based on the following aims:

– control of the combustion process that is responsible for heating the installation (by controlling the flow of air and fuel to maintain combustion);

– control of the parameters in the air heating process that will supply the blast furnace (by controlling the air flow at the entrance to the cowper and the temperature at the entrance to the blast furnace).

Combustion control includes closed-loop systems for the fuel flow rate (FRC-1), which complies with the setpoint of 98,000 m^3/h and for the combustion air flow

(FRC-2), which respects the setpoint with the value of 49,000 m³/h. Both control systems supply the thermal energy required to heat the air in the cowper.

For the (FRC-1) system, a first-order model $(\hat{b}_1,\hat{a}_1)$ was evaluated based on the technological values acquired during the evolution of the process,

$$\begin{aligned} \hat{b}_1 &= 0.19033\,z^{-1} \\ \hat{a}_1 &= 1-0.90484\,z^{-1} \end{aligned} \qquad [8.1]$$

and a digital polynomial algorithm ensuring fast aperiodic dynamics for tracking and regulation was calculated. The control algorithm was calculated using the pole placement method (see Chapter 7):

$$\begin{aligned} R_1 &= 0.0956-0.856z^{-1} \\ S_1 &= 0.4758-0.4758z^{-1} \\ T_1 &= 1-1.809z^{-1}+0.819z^{-2} \end{aligned} \qquad [8.2]$$

For the (FRC-2) system, the $(\hat{b}_2,\hat{a}_2)$ model was estimated using the same procedure:

$$\begin{aligned} \hat{b}_2 &= 0.1903\,z^{-1} \\ \hat{a}_2 &= 1-0.904\,z^{-1}, \end{aligned} \qquad [8.3]$$

and the corresponding control algorithm:

$$\begin{aligned} R_2 &= 0.5907-0.4731\,z^{-1} \\ S_2 &= 0.1903-0.1903\,z^{-1} \\ T_2 &= 1-1.314094\,z^{-1}+0.43171\,z^{-2} \end{aligned} \qquad [8.4]$$

To control the heating of air in a cowper, the following closed-loop systems are used: (FRC-3) for the air flow that supplies the air heating system (cowper) at a setpoint of 50,000 m³/h and (TRC-4) for the air temperature that feeds the blast furnace, which complies with the setpoint of 1,200°C.

For the (FRC-3) system, the second-order model $(\hat{b}_3,\hat{a}_3)$ was estimated:

$$\begin{aligned} \hat{b}_3 &= 0.06777z^{-1}+0.05188\,z^{-2} \\ \hat{a}_3 &= 1-1.3299z^{-1}+4.49258\,z^{-2} \end{aligned} \qquad [8.5]$$

and the corresponding digital control algorithm was obtained:

$$\begin{aligned} R_3 &= 8.35702 - 11.111503\, z^{-1} + 3.754475\, z^{-2} \\ S_3 &= 1 - 0.5663\, z^{-1} - 0.4336 z^{-2} \\ T_3 &= 8.35702 - 11.111503\, z^{-1} + 3.754475\, z^{-2} \end{aligned} \qquad [8.6]$$

The second-order dynamic model $(\hat{b}_4, \hat{a}_4)$ was evaluated and validated by an experimental identification procedure (RLS) for the (TRC-4) system:

$$\begin{aligned} \hat{b}_4 &= 0.00123 + 0.000139 z^{-1} \\ \hat{a}_4 &= 1 - 1.37198 z^{-1} + 0.37623 z^{-2} \end{aligned} \qquad [8.7]$$

The corresponding digital controller is composed of:

$$\begin{aligned} R_4 &= 1.76889 - 2.03422\, z^{-1} + 0.51875\, z^{-2} \\ S_4 &= 0.00123 + 0.000605\, z^{-1} - 0.001643\, z^{-2} \\ T_4 &= 1 - 0.99317 z^{-1} + 0.24660 z^{-2} \end{aligned} \qquad [8.8]$$

which ensures the imposed performances.

The three components of the combustion flow rate are calculated from a prespecified energy consumption recipe that is provided by the closed-loop systems: (FRC-5) for coke flow rate, (FRC-6) for CH_4 flow rate and (FRC-7) for blast furnace gas flow. We have calculated:

– the dynamic model $(\hat{b}_5, \hat{a}_5)$ of system (FRC-5):

$$\begin{aligned} \hat{b}_5 &= 0.03807 z^{-1} \\ \hat{a}_5 &= 1 - 0.9084 z^{-1} \end{aligned} \qquad [8.9]$$

and the corresponding control algorithm,

$$\begin{aligned} R_5 &= 0.27 - 0.23387 z^{-1} \\ S_5 &= 0.038065 - 0.038065\, z^{-1} \\ T_5 &= 1 - 1.63476\, z^{-1} + 0.67099\, z^{-2} \end{aligned} \qquad [8.10]$$

– model $(\hat{b}_6, \hat{a}_6)$ for system (FRC-6) with

$$\begin{aligned} \hat{b}_6 &= 0.0314z^{-1} \\ \hat{a}_6 &= 1 - 0.90484z^{-1} \end{aligned} \qquad [8.11]$$

and the control algorithm

$$\begin{aligned} R_6 &= 0.27 - 0.233846z^{-1} \\ S_6 &= 0.0314 - 0.0314z^{-1} \\ T_6 &= 1 - 1.63476\,z^{-1} + 0.670991\,z^{-2} \end{aligned} \qquad [8.12]$$

– model $(\hat{b}_7, \hat{a}_7)$ of system (FRC-7) with

$$\begin{aligned} \hat{b}_7 &= 0.47581\,z^{-1} \\ \hat{a}_7 &= 1 - 0.90484z^{-1} \end{aligned} \qquad [8.13]$$

and the polynomials of the control algorithm

$$\begin{aligned} R_7 &= 0.095682 - 0.85697z^{-1} \\ S_7 &= 0.475813 - 0.475813z^{-1} \\ T_7 &= 1 - 1.809155\,z^{-1} + 0.819142\,z^{-2} \end{aligned} \; . \qquad [8.14]$$

Before implementing the real-time control systems, the nominal performances of the closed-loop systems were tested in simulation. Some improvements were made to the system design using adaptive and robust strategies to keep performance in real time [LAN 97, DAU 04, POP 06].

8.1.2. *Optimization of the combustion process*

In terms of optimization, a multivariable nonlinear decision model was associated with the process and identified accordingly. The output of this model is the concentration of oxygen in the residual combustion gas, which shows the quality of the combustion process; the inputs are the flow rate of the fuel mixture and the air flow rate for combustion. The optimization problem was built and finally solved using a direct random search method, COMPLEX [POP 06, BOR 13a, STE 14].

For our case, the concentration of oxygen $z\left(\%O_2\right)$ was chosen as the quality variable (thus defining the optimization criterion), which depends on the inputs: combustion gas flow rate y_1 and combustion air flow rate y_2.

The main idea is to find the optimal solution for the combustion process that also meets the restrictive technological operating conditions.

The quasi-stationary decision model for the combustion process $\hat{z}\left(\%O_2\right)=f\left(y_1,y_2\right)$ was evaluated and, subsequently, the functions that express the technological constraints for concentration $\hat{z}_1$ of *CO*, dome temperature $\hat{z}_2$ of the installation, and temperature $\hat{z}_3$ of the residual gas all depend on the same variables y_1 and y_2, and were identified as:

$$\begin{aligned}
&\hat{z}_1\left(\%CO\right)=f_1\left(y_1,y_2\right)\\
&\hat{z}_2\left(T_{cowpercupola}\right)=f_2\left(y_1,y_2\right)\\
&\hat{z}_3\left(T_{flowgas}\right)=f_3\left(y_1,y_2\right)
\end{aligned} \quad [8.15]$$

These nonlinear functions (models) were evaluated using the LS experimental identification technique.

The experimental protocol concentrates on the data acquisition procedure during the first heating interval of the installation, with an imposed duration (resolution of 128 observations with a 2-s acquisition period).

With the measurements collected in real time, the following models were estimated:

$$\begin{aligned}
&\hat{z}=-9.665+0.229y_1-0.0009y_1^2+0.010y_2\\
&\hat{z}_1=-4282.875-21.566y_1-0.077y_1^2+21.500y_2\\
&\hat{z}_2=1277.613+0.001y_1^2-0.387y_2\\
&\hat{z}_3=499.161926-0.002147y_1-3.49945y_2
\end{aligned} \quad [8.16]$$

The first equation represents the decision model and the set, $\hat{z}_1, \hat{z}_2, \hat{z}_3$, represents the models of the technological constraints of the process.

The nonlinear parametric optimization problem is expressed as:

$$\min\{I = -9.665 + 0.229y_1 - 0.0009y_1^2 + 0.010y_2\} \qquad [8.17]$$

by considering all the constraints:

$$\begin{aligned} &0 \le \hat{z}_1 \le 450\,ppm \\ &0 \le \hat{z}_2 \le 1300^0 C \\ &0 \le \hat{z}_3 \le 340^0 C \\ &96.309 \le y_1 \le 102.452 \\ &46.602 \le y_2 \le 57.992 \end{aligned} \qquad [8.18]$$

The solution to this problem, which is calculated using the direct digital method BOX, determines the optimal point of operation of the combustion process:

– $y_1^* = 97469.85\,m^3/h$ – optimal flow of combustible gas;

– $y_2^* = 47804.16\,m^3/h$ – optimal flow of combustion air.

This operating point guarantees the following minimum value of O_2 in the residual gas:

$$z^*_{\min}(\%O_2) = 4.65\% \qquad [8.19]$$

Using the approach presented in this case study, methane fuel consumption was reduced by 7.4 % from the average level of consumption.

The values of the constraint variables calculated at this operating point are as follows:

$$\begin{aligned} &\overset{*}{z}_1(\%CO) = 415.73\,ppm \\ &\overset{*}{z}_2(T_{cupola}) = 1273.25^o C \\ &\overset{*}{z}_3(T_{flowgas}) = 311.47^o C \end{aligned} \qquad [8.20]$$

The optimal decision $\left(y_1^*, y_2^*\right)$ is automatically taken as setpoints $\left(r_1^* = y_1^*, r_2^* = y_2^*\right)$ for the execution level, which leads the combustion process to the optimal point of operation through closed-loop systems.

The optimization problem [8.17 and 8.18] is reconstructed and solved once the decision model $\hat{z}$ is updated. This application has been put into place and it controls a blast furnace heating setup on a steel plant in south-eastern Europe [POP 06, FIL 09].

8.2. Control and optimization of an ethylene installation

The chemical industry is still a fertile domain for automatic control of technological processes, and as a result some of the most representative applications find their place in this sector. In recent decades, the petrochemical industry has undergone unprecedented development by improving equipment and production lines while also contributing to the expansion of production capacity.

Today, the demand for ethylene exceeds 125 million tons per year with a yearly growth trend of 3.5%. It is expected that the average capacity of factories, worth 300 KTA in the 80s and over 1,000 KTA today, will grow beyond 2,000 KTA in the near future.

Pyrolysis is a petrochemical reaction for the thermal decomposition of complex organic matter (petrol, ethane) into other valuable products: ethylene, propylene, butadiene, benzene and xylenes, which form the basis of the entire petrochemical industry. Organic matter consists of chains of molecules. Heat breaks these chains and produces simpler organic structures. The chemical reactions between organic matter and steam occur in a low-oxygen atmosphere in order to avoid dangerous explosions.

Therefore, in pyrolysis furnaces, a complex reaction takes place between the organic component and water (in vapor phase) at high temperatures (850°C) and low pressure (3–4 atm) and will generate a complex mixture of chemicals in gas phase [CAL 79, TER 89].

Ethylene is the main product and represents one of the most important raw materials in the petrochemical industry. It is considered to be the backbone of the petrochemical industry and is regarded as the product with the largest production volume as well as being the starting point for most derived products, including polyethylenes and glycols, which consume about 85% of ethylene production.

There are two working areas in the reactor: the convection zone in which the necessary raw material and steam are mixed, where the reaction will start at a temperature of about 500°C, and the radiation zone in which the chemical reactions take place at temperatures between 820°C and 860°C. The pressure in the reactor remains between 3 and 4 atm.

The heat required for the reaction is provided by a battery of burners, supplied with methane fuel.

The product of the pyrolysis reaction is a mixture of chemical components from which the main product, ethylene, is extracted.

After the ethylene is recovered, a complex sequence of procedures for separating the various chemical components takes place.

Operating the pyrolysis reactor at a nominal working point ensures a correct ratio between the gasoline and steam, and the necessary temperature and pressure conditions for the pyrolysis reaction. To achieve greater chemical efficiency, the nominal control solution must be improved and an optimum operating point must be found for the reactor.

8.2.1. *Automation solution and designing the control algorithms*

The control solution for the pyrolysis installation is designed for two important objectives:

– to guarantee a correct ratio between the reactants at the reactor inlet. These reactants participate in the reaction through control of the gas flow and the steam flow rate;

– to ensure the necessary conditions for the chemical reaction by controlling the temperature in the middle section of the reactor, which must remain around a value of 850°C, and by controlling the pressure inside the reactor, which must be maintained at a value of 4 atm.

The control models for reactant flow control systems were evaluated using an analytical approach (balance equations), and for the temperature and pressure of the pyrolysis reaction the dynamic models were identified by RLS experimental techniques.

The design of digital control systems using the pole placement method is carried out using RST polynomial algorithms. Before implementing the control, we simulated the nominal performances in closed-loop systems. After an analysis of the robustness of the systems, the control algorithm is improved in order to keep the performances in real time during the physical process exploitation [FRA 72, LAN 97, POP 06, POP 13].

8.2.1.1. *Controlling the ratio of gasoline and steam flows*

The first objective is to properly calibrate the quantities introduced into the reactor that participate in chemical reactions. This is done by controlling the gasoline and steam flow rates.

Initially, analytical models are calculated (see Chapter 7) based on mechanical fluid flow equations. These models are expressed continuously by the transfer functions:

$$G_{F1}(s) = \frac{0.24}{(8s+1)(0.67s+1)} \quad [8.21]$$

for the flow of gasoline and,

$$G_{F2}(s) = \frac{0.21}{(4s+1)(0.2s+1)} \quad [8.22]$$

for steam flow.

The PI algorithms are calculated by placing the imposed poles for closed-loop systems. The corresponding transfer functions are described by the following equations:

$$G_{R1}(s) = 11.11\left(1+\frac{1}{8s}\right). \qquad G_{R2}(s) = 6.34\left(1+\frac{1}{4s}\right) \quad [8.23]$$

Using the Euler technique with a sampling period of 0.1 s, we obtained the discrete equivalent representation in [8.24] for both algorithms from [8.23]:

$$G_{R1}(z^{-1}) = \frac{11.24-11.11z^{-1}}{1-z^{-1}}. \qquad G_{R2}(z^{-1}) = \frac{7.92-6.34z^{-1}}{1-z^{-1}} \quad [8.24]$$

The imposed tracking and regulation performances for the two closed-loop systems were validated in simulation using Matlab-Simulink software. These results are presented in Figure 8.1 showing their performances.

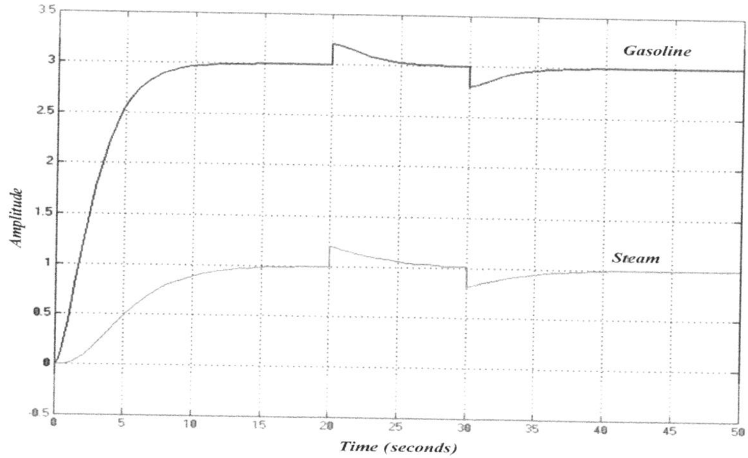

Figure 8.1. *Response of gasoline and steam flow control systems*

8.2.1.2. *Controlling chemical reaction parameters*

The second objective of the proposed automation solution is to ensure that the required temperature and pressure conditions for the chemical reaction are achieved by controlling the temperature and pressure inside the reactor.

Mathematical models were evaluated by an experimental identification procedure using the RLS method. During the design stage of the control, the algorithms that provide performance for closed-loop systems were calculated using the discrete pole placement technique.

A second-order numerical model was identified and validated for the (TRC-3) temperature control system from an experimental data acquisition and processing program, and is presented in [8.25]:

$$G_{P3}(z^{-1}) = \frac{0.04711z^{-1}}{1-1.61402z^{-1}+0.65344z^{-2}}. \qquad [8.25]$$

For the (TRC-3) system, the calculated digital control algorithm is given by relationship [8.26]:

$$\left[\begin{array}{l} R_3(z^{-1}) = 36.45 - 52.01z^{-1} + 18.98z^{-2} \\ S_3(z^{-1}) = 1 + 0.36z^{-1} - 1.36z^{-2} \\ T_3(z^{-1}) = 21.22 - 26.42z^{-1} + 8.63z^{-2} \end{array}\right. . \qquad [8.26]$$

The model identified for the (PRC-4) pressure control system is presented in [8.27]:

$$G_{P4}(z^{-1}) = \frac{0.00597z^{-1}}{1-1.68364z^{-1}+0.70730z^{-2}}. \quad [8.27]$$

The digital RST control algorithm for the (PRC-4) control system was evaluated using the same pole placement procedure, with independent objectives for tracking and regulation, being expressed by the polynomials:

$$\begin{cases} R_4(z^{-1}) = 241.9 - 372.9z^{-1} + 145.3z^{-2} \\ S_4(z^{-1}) = 1 + 0.22z^{-1} - 1.22z^{-2} \\ T_4(z^{-1}) = 167.5 - 243.9z^{-1} + 90.8z^{-2} \end{cases}. \quad [8.28]$$

Figure 8.2 simulates the response and control algorithm of the closed-loop system (TRC-3):

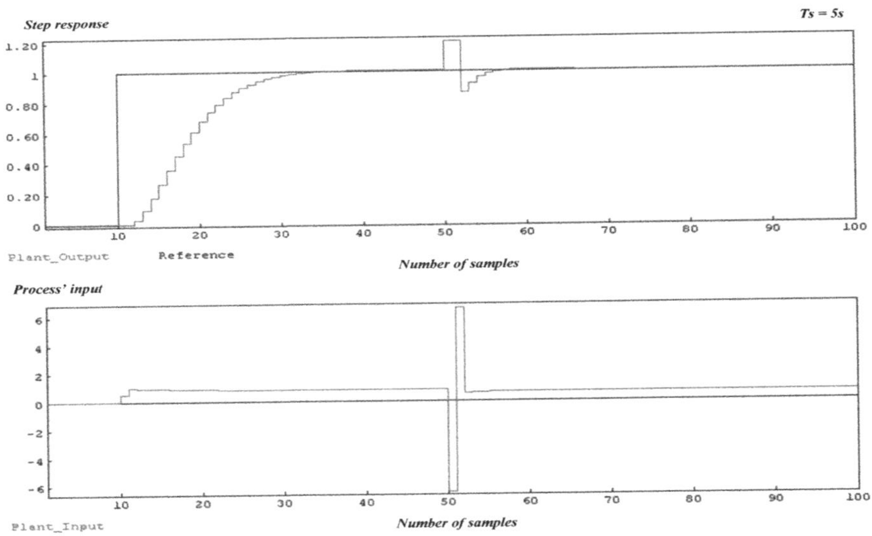

Figure 8.2. *Response of the (TRC-3) temperature control system. For a color version of this figure, see http://www.iste.co.uk/popescu/controldesign.zip*

Figure 8.3 shows the tracking and regulation dynamics of the (PRC-4) system and control algorithm of the closed-loop system.

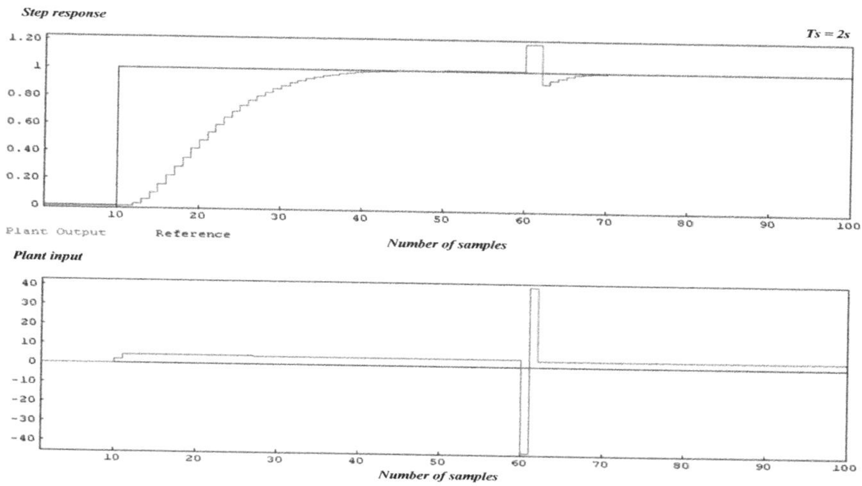

Figure 8.3. *Response of the pressure control system. For a color version of this figure, see http://www.iste.co.uk/popescu/controldesign.zip*

The ultimate benefit of advanced automatic control is to improve the dynamics of the control algorithm in order to maintain the nominal performance as the performance on the physical process. Before implementing the control on a microcontroller, we simulated the nominal performances of closed-loop systems and made some improvements to the control algorithms using a robust strategy to keep the performance on the installation in real time [LAN 97, POP 06].

A robust analysis is performed on the (TRC-3) and (PRC-4) systems, which ensures the conditions of kinetics in the pyrolysis reactions. This analysis is based on the calculation of a robustness modulus ΔM and on the frequency characteristic of the perturbation-output sensitivity function of the system.

The modulus margin $|\Delta M(j\omega)|$ measures the minimum distance between the critical point [–1, 0*j*] and the Nyquist locus of the open-loop system.

For the control system in consideration with a RST control algorithm, the sensitivity function is expressed by:

$$S_{py}(z^{-1}) = \frac{\hat{a}(z^{-1})S(z^{-1})}{\hat{a}(z^{-1})S(z^{-1}) + \hat{b}(z^{-1})R(z^{-1})}, \qquad [8.29]$$

where $\hat{a}(z^{-1})$ and $\hat{b}(z^{-1})$ are the polynomials of the identified model and $R(z^{-1})$ and $S(z^{-1})$ are the polynomials of the calculated control algorithm.

The relationship between the two robustness indicators is known:

$$|\Delta M(j\omega)| = \left|S_{py}^{-1}(j\omega)\right|_{\min} = \frac{1}{\left|S_{py}(j\omega)\right|_{\max}} \qquad [8.30]$$

An accepted robustness is expressed in practice by a module that is greater than 0.5, or by a maximum value for a sensitivity function of less than 6 dB.

The frequency characteristic of the sensitivity function for the nominal control system, as presented in equations [8.28], shows that the robustness indicators are not respected (Figure 8.4).

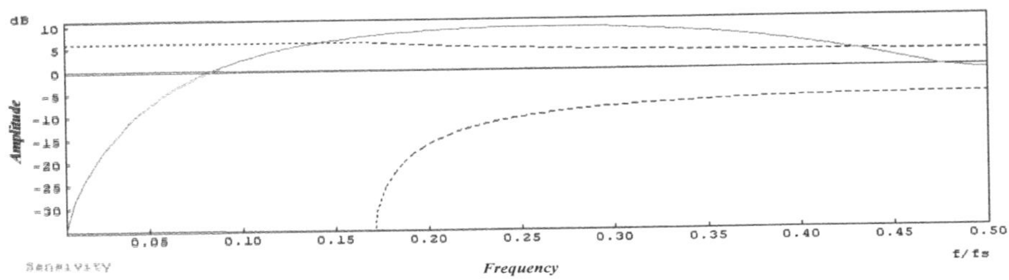

Figure 8.4. *Sensitivity function of the nominal temperature control system. For a color version of this figure, see http://www.iste.co.uk/popescu/controldesign.zip*

To increase the robustness of the system, a control algorithm correction was carried out and another temperature control algorithm (TRC-3) was calculated:

$$\begin{cases} R_3(z^{-1}) = 3.42 - 5.53z^{-1} + 2.24z^{-2} \\ S_3(z^{-1}) = 1 - 1.24z^{-1} + 0.24z^{-2} \\ T_3(z^{-1}) = 21.22 - 60.68z^{-1} + 65.15z^{-2} - 31.19z^{-3} + 5.63z^{-4} \end{cases} . \qquad [8.31]$$

The robust algorithm of the (TRC-3) system corrected the sensitivity function, keeping it below the imposed 6 dB limit, and tracking performance was retained by the enhanced control dynamics (see Figure 8.5).

For the (PRC-4) pressure control system, a similar correction was applied and the calculated robust polynomial algorithm corrected the sensitivity function in addition to keeping it below the 6 dB limit (see Figure 8.6) through the following improved control dynamics:

$$\begin{cases} R_4(z^{-1}) = 14.34 - 24.15z^{-1} + 10.14z^{-2} \\ S_4(z^{-1}) = 1 - 1.45z^{-1} + 0.45z^{-2} \\ T_4(z^{-1}) = 167.5 - 526.6z^{-1} + 620.1z^{-2} - 325.5z^{-3} + 64.24z^{-4} \end{cases} \quad [8.32]$$

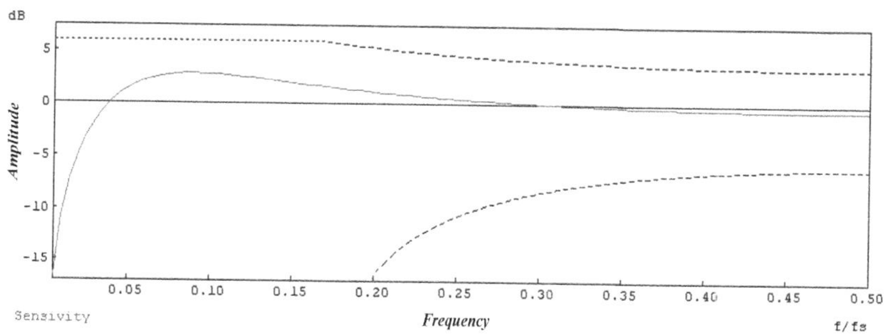

Figure 8.5. *Sensitivity function of the robust temperature control system. For a color version of this figure, see http://www.iste.co.uk/popescu/controldesign.zip*

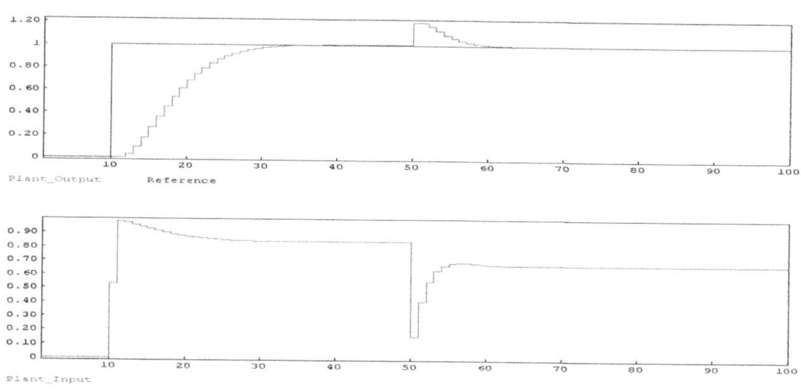

Figure 8.6. *Response of the robust temperature control system. For a color version of this figure, see http://www.iste.co.uk/popescu/controldesign.zip*

8.2.2. *Optimizing the pyrolysis process*

The optimization problem ensures the optimal operating point of the pyrolysis reactor. In our case study, the aim of optimization was to maximize the ethylene concentration (z), and thus the conversion efficiency of the gasoline/raw material into the ethylene/useful product by calculating a decision for new setpoints of closed-loop systems. To solve this problem, a certain amount of data was acquired for the inputs: gas flow rate (y_1), steam flow rate (y_2), reaction pressure (y_3) and reactor temperature (y_4) and for output (z), the quality variable that represents the concentration of ethylene in the reaction product.

Based on a well-chosen data set, we estimated the nonlinear, multivariable chemical process model with output $\hat{z}$ through the following structure:

$$\hat{z} = \hat{a}_0 + \hat{a}_1 y_1 + \hat{a}_2 + \frac{1}{y_2} + \hat{a}_3 y_3 + \hat{a}_4 y_4^2 \quad [8.33]$$

where

$$\hat{a}^T = \begin{bmatrix} \hat{a}_0 & \hat{a}_1 & \hat{a}_2 & \hat{a}_3 & \hat{a}_4 \end{bmatrix} \quad [8.34]$$

is the vector of unknown parameters of the model, which was estimated using the LS method on the acquired experimental data set.

The optimization problem was then constructed in order to maximize the main product, ethylene, as follows:

$$\max \hat{z}(y) \quad [8.35]$$

with the constraints:

$$\begin{aligned} &1000 \le y_1 \le 1600[m^3/h] \\ &430 \le y_2 \le 540[m^3/h] \\ &3.3 \le y_3 \le 4.5[barr] \\ &820 \le y_4 \le 860[^0C] \end{aligned} \quad [8.36]$$

The results obtained from Matlab, using a direct search method (BOX), led to the following values:

$y^*_1 = 1559.6\ [\mathrm{m}^3/\mathrm{h}]$, $y^*_2 = 430.03\ [\mathrm{m}^3/\mathrm{h}]$, $y^*_3 = 3.3\ [barr]$

$y^*_4 = 859.85\ [^0\mathrm{C}]$.

The proportion of ethylene in the final product from the pyrolysis reaction gave the optimum value: $\hat{z}^* = 34.16\%$.

The optimal decision $\left(y^*_1, y^*_2, y^*_3, y^*_4\right)$ of this optimization level is used as the setpoints for the closed-loop systems of the control level and guides the process evolution toward the suitable optimal point.

After applying this approach in real time, the concentration of ethylene increased by 4.2% relative to the average value for a significant petrochemical platform, according to [POP 06, FIL 09, BOR 13, STE 14].

8.3. Digital control of a thermoenergy plant

The recommended solution for increasing the operating efficiency of heating systems today is to introduce, on a large scale, high-performance automation systems control that pilot modern technologies in facilities that provide the thermal energy.

The main objective of this case study is to design a digital development structure that serves existing installations at a thermal energy distribution point within a large urban heating network. The proposed control configuration is a flexible hardware–software structure capable of adapting to seasonal, real-world operating conditions and user requirements. The control systems of the relevant thermal installations ensure the acquisition and regulation of parameters that are technologically important in the process (household water temperature, temperature of the secondary heating agent, pressure drop at the thermal point) providing the heat transfer operations from agent to products.

Implementing a high-performance digital development system requires knowledge of technologies, operational and managerial resources, integration of theoretical and practical knowledge of systems (methods, algorithms and software products) and equipment (process interfaces, microcontrollers, programmable logic controllers, translators and intelligent actuators).

In this case study, we developed a solution over two stages: the first involved the design of the digital control level and the second was to design a supervisory level to ensure that the agent-product thermal transfer was optimized.

8.3.1. *Solving the problem of automation of a thermal operating point*

The automation solution is composed of a series–parallel technological configuration that exists at a thermal point in the urban heating network. This configuration is used to ensure an efficient heat transfer between the agent-product pair for domestic heating and domestic hot water [POP 06, TUD 13].

In principle, centralized heating systems have heat transfer processes that present a high degree of complexity and pose many problems in their operation.

From the premises presented below, we propose an efficient solution that provides the following functions: monitoring information at a thermal point by measuring and acquiring data throughout the process, controlling parameters of technological importance at the thermal point and optimizing energy transfer.

The automatic control function targets the technological parameters of the facility for which the following control systems are provided: (TRC-1), the temperature of domestic hot water for which the operating variable is the primary thermal agent (at the inlet of the installation); (TRC-2), the temperature of the secondary heating agent for which the operating variable is a partial flow rate from the thermal agent; (DPRC-3), the pressure drop at the thermal point, with the operating variable being the output flow rate of the thermal agent. Besides these process parameters, there are also flow and temperature of the thermal agent, and input and output energy at the thermal point, which are measured by the data acquisition system.

These control systems ensure operation of the facility at a nominal point at the values imposed on the operator by the rules of operation.

To design these control systems, mathematical models were estimated using the RLS identification method. The control algorithms, which ensure dynamic performance, were calculated using an appropriate digital design technique [LAN 97, DAU 04, POP 06, TUD 13].

The performance requirements for closed-loop systems were validated in simulation by dedicated software.

The discrete dynamic model of the process was estimated for the (TRC-1) system by identification through the RLS technique:

$$G_f(z^{-1}) = \frac{0.21z^{-1}}{1-0.88z^{-1}}z^{-1}. \tag{8.37}$$

Based on the estimated model, a digital polynomial algorithm (the R and S polynomials) was calculated to ensure the imposed process performances occurred.

Tracking dynamics are specified by the $(\hat{b}_m, \hat{a}_m)$ model, which generated the imposed trajectory:

$$\begin{cases} \hat{b}_m(z^{-1}) = 0.02 + 0.02\,z^{-1} \\ \hat{a}_m(z^{-1}) = 1 - 1.56\,z^{-1} + 0.61\,z^{-2} \end{cases}. \tag{8.38}$$

This trajectory will be reproduced at the output of the closed-loop system by the available T (z^{-1}) polynomial. The regulation performance was imposed by the dynamic system obtained by discretizing a second-order system:

$$G_0(s) = \frac{\omega_n^2}{s^2 + 2\zeta\omega_n + \omega_n^2}, \tag{8.39}$$

with $\omega_n = 0.0042\ rad/\sec$ and $\zeta = 0.95$, which ensures the desired pole placement for the closed-loop system.

The general algorithm for the calculated RST control is expressed by the following polynomial components:

$$\begin{cases} R(z^{-1}) = 10.68 - 6.80\,z^{-1} \\ S(z^{-1}) = 1 + 0.65\,z^{-1} - 1.65\,z^{-2} \\ T(z^{-1}) = 4.65 - 1.03\,z^{-1} + 0.26\,z^{-2} \end{cases}. \tag{8.40}$$

The response of the closed-loop (TRC-1) system simulation, which provides tracking and regulation performance as well as control algorithm dynamics, is shown in Figure 8.7.

For the (TRC-2) system, a first-order model with an important delay of the process was evaluated using an experimental RLS identification technique,

$$G_f(z^{-1}) = \frac{0.47\,z^{-1}}{1 - 0.86\,z^{-1}}\,z^{-3}. \tag{8.41}$$

An RST algorithm that ensures imposed tracking and control performance was calculated using the same procedure.

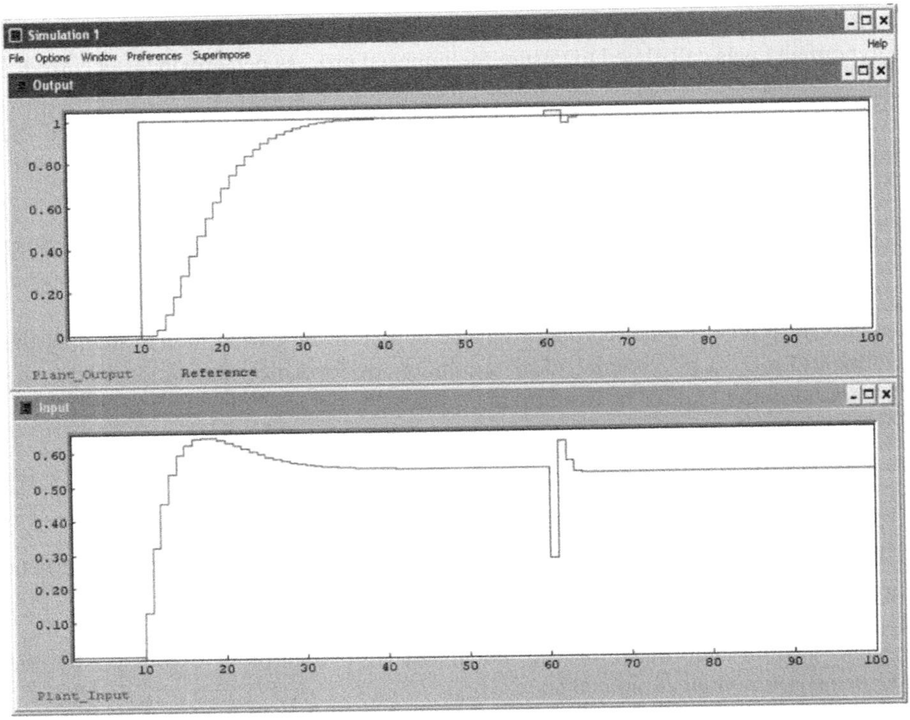

Figure 8.7. *Response of the closed-loop (TRC-1) system*

The tracking dynamics is imposed by the discrete model $(\hat{b}_m / \hat{a}_m)$ with:

$$\begin{cases} \hat{b}_m(z^{-1}) = 0.037 + 0.03\,z^{-1} \\ \hat{a}_m(z^{-1}) = 1 - 1.49z^{-1} + 0.56\,z^{-2} \end{cases}. \quad [8.42]$$

The output of model [8.42] gives the tracking dynamics that will be reproduced at the output of the closed-loop system by calculating the precompensator $T(z^{-1})$.

The regulation performance is fixed by the characteristic polynomial obtained by discretization of the second-order dynamic system,

$$G_0(s) = \frac{\omega_n^2}{s^2 + 2\zeta\omega_n + \omega_n^2} \quad [8.43]$$

with $\omega_n = 0.001\,rad/\sec$ and $\zeta = 0.95$, which ensures the desired pole placement.

RST control for a closed-loop system includes the following polynomial components:

$$\begin{cases} R(z^{-1}) = 0.26 - 0.22\,z^{-1} \\ S(z^{-1}) = 1 - 0.89\,z^{-1} + 0.006\,z^{-2} + 0.006\,z^{-3} - 0.12\,z^{-4} \\ T(z^{-1}) = 2.11 - 3.71\,z^{-1} + 1.64\,z^{-2} \end{cases}. \qquad [8.44]$$

The response of the closed loop (TRC-2) system and the control dynamics in simulation are presented in Figure 8.8.

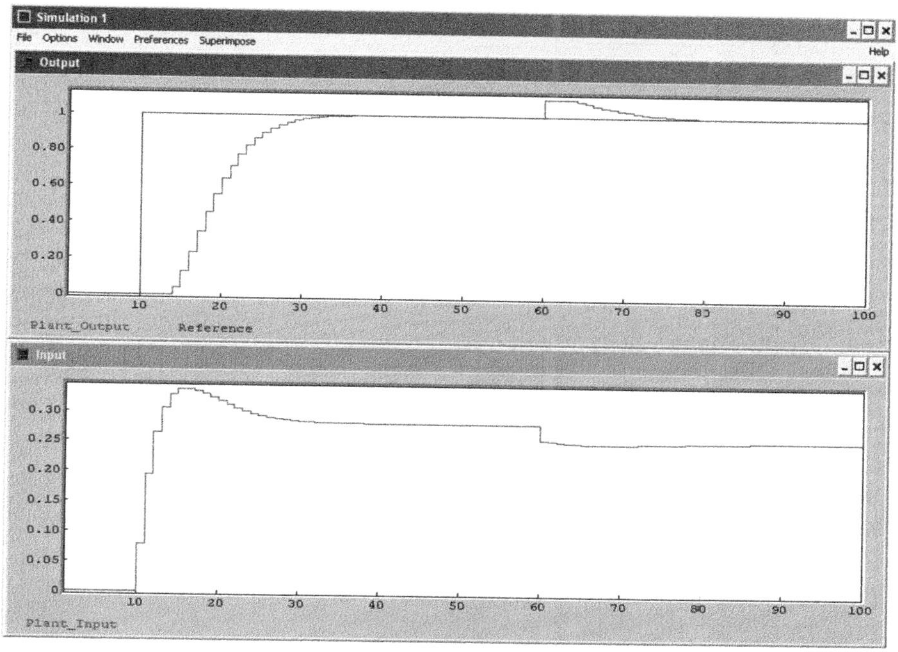

Figure 8.8. *Response of the closed-loop (TRC-2) system*

The discrete process model for the (DPRC-3) system was also estimated by experimental identification:

$$G_f(z^{-1}) = \frac{0.019\,z^{-1} + 0.015\,z^{-2}}{1 - 1.45\,z^{-1} + 0.51\,z^{-2}}. \qquad [8.45]$$

An RST algorithm that ensures compulsory tracking and control performance was calculated.

The tracking dynamic is specified using the following discrete model:

$$\begin{cases} \hat{b}_m(z^{-1}) = 0.037 + 0.03\,z^{-1} \\ \hat{a}_m(z^{-1}) = 1 - 1.49z^{-1} + 0.56\,z^{-2} \end{cases} . \qquad [8.46]$$

The control dynamic is obtained as before through the discretization of a second-order continuous system:

$$G_0(s) = \frac{\omega_n^2}{s^2 + 2\zeta\omega_n + \omega_n^2} \qquad [8.47]$$

with $\omega_n = 0.002\, rad / \sec$ and $\zeta = 0.9$, which are values that define the poles of the closed-loop system.

The RST controller is represented by the set of polynomials:

$$\begin{cases} R(z^{-1}) = 23.36 - 35.39\,z^{-1} + 13.44\,z^{-2} \\ S(z^{-1}) = 1 - 0.58\,z^{-1} - 0.41\,z^{-2} \\ T(z^{-1}) = 28.23 - 44.82\,z^{-1} + 18\,z^{-2} \end{cases} . \qquad [8.48]$$

The response of the closed-loop (DPRC-3) system, for tracking and control, and the control algorithm dynamics are presented (in simulation) in Figure 8.9.

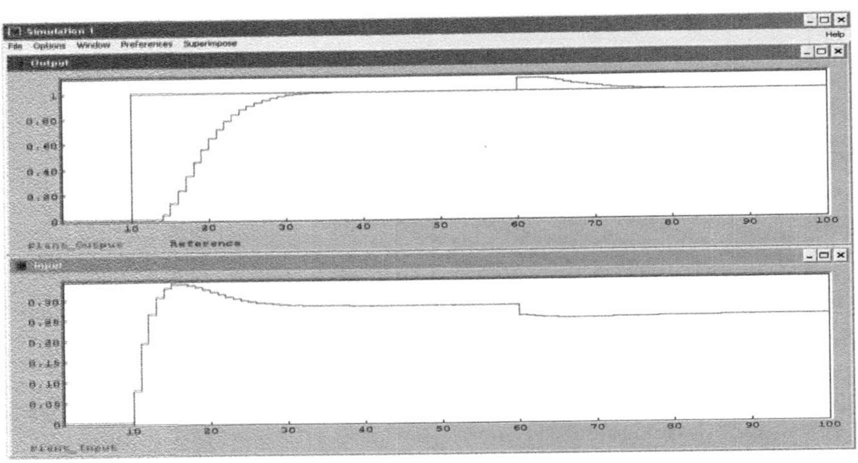

Figure 8.9. *Response of the closed-loop system, DPRC-3*

8.3.2. *Optimization of thermal transfer and agent product*

The supervisory level, which is integrated into the hierarchical control process, achieves the optimization function of the thermoenergy transfer process.

As a quality variable z, we propose the temperature of the thermal agent on the output of the thermal point and the outputs are as follows: y_1 is the temperature of the domestic water, y_2 is the heating temperature and y_3 is the pressure drop of the thermal agent.

We consider that the efficient thermal-product agent heat transfer is ensured, if the thermal residue evacuated by the output thermal agent is minimal.

The structure of the nonlinear multivariable model is:

$$\hat{z} = \hat{a}_0 + \hat{a}_1 y_1 + \hat{a}_2 y_2 + \hat{a}_3 y_3^{\ 2} \qquad [8.49]$$

The vector of the model parameters of the form $\hat{a} = [\hat{a}_0 \quad \hat{a}_1 \quad \hat{a}_2 \quad \hat{a}_3]^T$ is determined by the LS method using appropriately sized digital data files. In this sense, a set of files with the necessary input–output data was acquired.

The parameters of the estimated model are:

$$\begin{cases} \hat{a}_0 = 74.85 \\ \hat{a}_1 = -2.15 \\ \hat{a}_2 = -38.06 \\ \hat{a}_3 = 5.68 \end{cases} \qquad [8.50]$$

The criterion function is constructed as follows:

$$I(y) = y_2 \cdot \hat{z}(y), \qquad [8.51]$$

where $y = [y_1, y_2, y_3]$ is the vector of the input variables.

The constraints below are interpreted as the lower and upper limits of the measured values:

$$\begin{aligned} &42.67\ {}^0C \le y_1 \le 65.18\ {}^0C \\ &34.20\ {}^0C \le y_2 \le 47.62\ {}^0C\ . \\ &2.67\ barr \le y_3 \le 3.98\ barr \end{aligned} \qquad [8.52]$$

The results of the optimization problem are as follows:

$$\begin{aligned} z_{\min} &= 30.98\ ^0C \\ y_{1\_\min} &= 63.09\ ^0C \\ y_{2\_\min} &= 35.44\ ^0C \\ y_{3\_\min} &= 2.67\ barr \end{aligned} \quad . \qquad [8.53]$$

The results obtained are within the imposed limits.

The solution $\left(y_1^* = 63.09\ ^0C \quad y_2^* = 35.44\ ^0C \quad y_3^* = 2.67\ barr\right)$ for the optimization problem represents the optimal decision for operation at the thermal point with maximized efficiency for heat transfer. For the measured data file, the mean value $z_{average} = 40.33^oC$ is calculated. Let us note that $z_{\min} < z_{\text{average}}$. The value of the quality indicator, which is calculated after the optimization operation, is $r = (z_{\min} - z_{\text{average}})/z_{\min} = 10.34\%$ and hence the conversion efficiency obtained is satisfactory.

The values calculated at the supervisory level are automatically transferred into the configuration of the control systems in the form of calculated (optimal) setpoints. The operations for identifying the decision model and the optimization model are repeated according to the seasonal operating routine of the facility under supervision [CAL 79, TER 90, POP 06, FIL 09, BOR 13].

8.4. Extremal control of a photovoltaic installation

In this case study, we design an extremal system for controlling the maximum power of a photovoltaic panel. The generic configuration of a solar platform is shown in Figure 8.10.

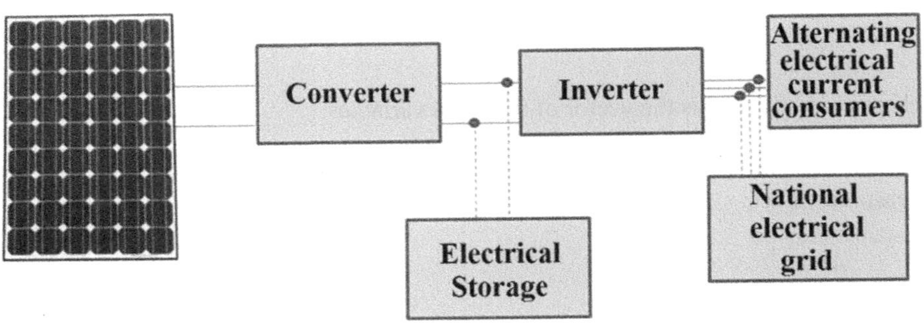

Figure 8.10. *Diagram of a photovoltaic platform*

Photovoltaic electricity is an intermittent source of energy; it is nonlinear and depends on many environmental parameters, such as solar radiation and temperature. This energy source must, therefore, be adapted to the dynamics of an adequate representation model. The power supplied by the photovoltaic panel depends on the operating point for the characteristic of the generated power. In fact, the photovoltaic panel is considered as a collection of PN junctions, which are the basis of photovoltaic cells. Many mathematical models have been developed to represent the behavior of photovoltaic cells.

The operation of a PN junction allows us to obtain an electrical model equivalent to the photovoltaic cell shown in Figure 8.11, an ideal model. It is the simplest model for the solar cell because it only takes the phenomenon of diffusion into account.

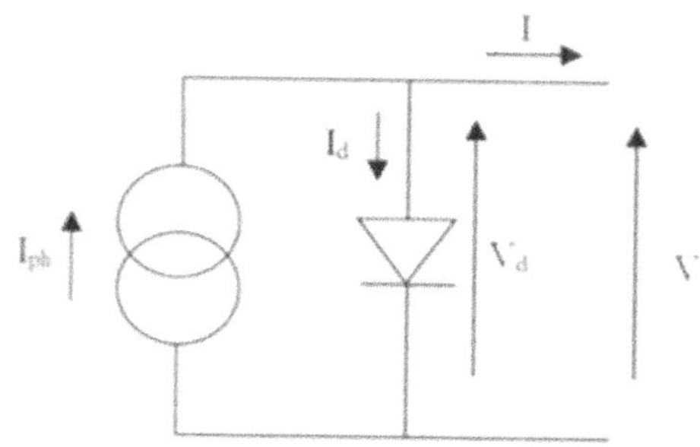

Figure 8.11. *Ideal model of a PV cell*

The associated electronic equations are as follows:

$$I = I_{ph} - I_d$$

$$I_d = I_0 \left(e^{\frac{V_d}{\alpha V_T}} - 1 \right)$$

and we therefore get the general formula for current:

$$I = I_{ph} - I_0 \left(e^{\frac{V_d}{\alpha V_T}} - 1 \right), \qquad [8.54]$$

where I is the current supplied by the cell, V is the voltage across the cell, I_0 is the reverse saturation current of the diode, I_{ph} is the short-circuit current outputed by the cell, a is the ideality factor of the diode and V_T is the thermodynamic potential with $V_T = \frac{k_b}{q}T$. The unknown parameters of the model are as follows: I_{ph}, I_0 and α.

This type of model has the advantage of containing a reduced number of parameters that are required for calculating the characteristic curve. However, it does not take environmental variations into account, so it cannot function in the real world.

The model with one diode (Figure 8.12) is improved by adding a resistor in parallel (noted R_{sh} or R_p) and a resistor in series (R_s) to the previous structure.

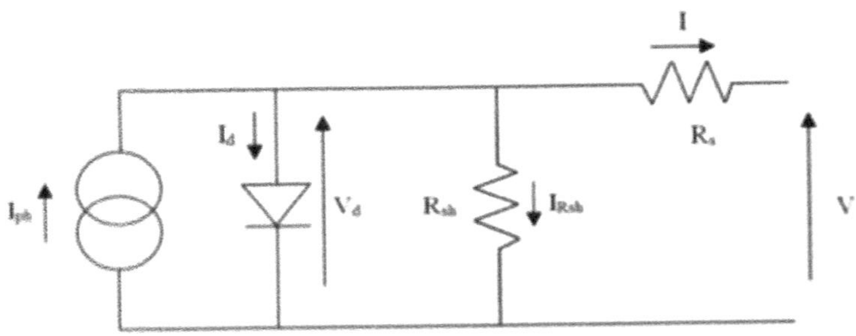

Figure 8.12. *Model of a PV cell with one diode*

It considers the ohmic losses in the cell and is described by the following equations:

$$
\begin{aligned}
I &= I_{ph} - I_d - I_{R_p} \\
I_d &= I_0\left(e^{\frac{V_d}{\alpha V_T}} - 1\right) \text{ with } V_d = V + IR_s \\
I_{R_s} &= \frac{V_d}{R_p} = \frac{V + IR_s}{R_p} \\
I &= I_{ph} - I_0\left(e^{\frac{V+IR_s}{\alpha V_T}} - 1\right) - \frac{V + IR_s}{R_p}
\end{aligned}
\qquad [8.55]
$$

The unknown parameters are as follows: I_{ph}, I_0, a, R_S and R_p.

This model is satisfactory and is even a reference for builders who rely on this model to provide the technical characteristics of their solar cells.

The two-diode model is said to be the closest to the actual behavior of the solar cell because it considers the charge transfer mechanism inside the cell. The additional diode makes it possible to reproduce the chemical recombination effects of electrons in the equivalent diagram (Figure 8.13).

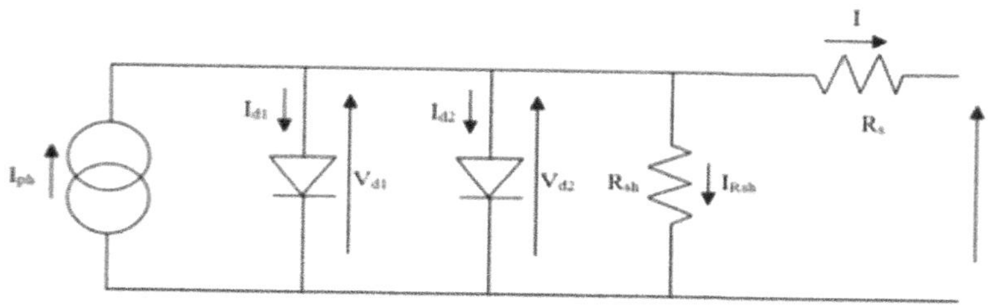

Figure 8.13. *A two-diode model of a PV cell*

$$I = I_{ph} - I_{01}\left(e^{\frac{V+IR_S}{\alpha_1 V_T}} - 1\right) - I_{02}\left(e^{\frac{V+IR_S}{\alpha_2 V_T}} - 1\right) - \frac{V+IR_S}{R_p} \quad [8.56]$$

In this figure, I is the current supplied by the cell, V is the voltage across the cell, I_{ph} is the short-circuit current delivered by the cell, I_{01} is the reverse saturation current of diode 1, I_{02} is the reverse saturation current of diode 2, a_1 is the ideality factor of diode 1, a_2 is the ideality factor of diode 2, V_T is the thermodynamic potential, R_s is the series resistance, and R_p is the parallel resistance.

The unknown parameters of the model are as follows: I_{ph}, I_{01}, I_{02}, a_1, a_2, R_S and R_p.

The performances of the photovoltaic module are estimated by the nonlinear characteristic curves $I = f(V)$ and $P = f(V)$. There are several methods for drawing the photovoltaic nonlinear characteristic curve. In particular, we used the Newton–Raphson method, which is based on a recursive digital technique for solving nonlinear equations of the type $f(x)=0$. Indeed, we seek the root (zero) of function f in a recurring manner by considering its first-order Taylor development. By applying this method to the two-diode model, the aim is to solve the following equation for different values of V such that $0 \le V \le V_{oc\_STC}$. We consider that,

$$I = I_{ph} - I_{01}\left(e^{\frac{V+IR_S}{\alpha_1 V_T}} - 1\right) - I_{02}\left(e^{\frac{V+IR_S}{\alpha_2 V_T}} - 1\right) - \frac{V+IR_S}{R_p} = 0. \qquad [8.57]$$

For this, let:

$$f(I) = I - I_{ph} + I_{01}\left(e^{\frac{V+IR_S}{\alpha_1 V_T}} - 1\right) + I_{02}\left(e^{\frac{V+IR_S}{\alpha_2 V_T}} - 1\right) + \frac{V+IR_S}{R_p} \qquad [8.58]$$

with $0 \le I \le I_{sc\_STC}$, and the Newton–Raphson method is applied to function f defined in [8.58].

The extremal profiles of the characteristics $I = f(V)$ and $P = f(V)$ of a photovoltaic cell are shown in Figures 8.14 and 8.15.

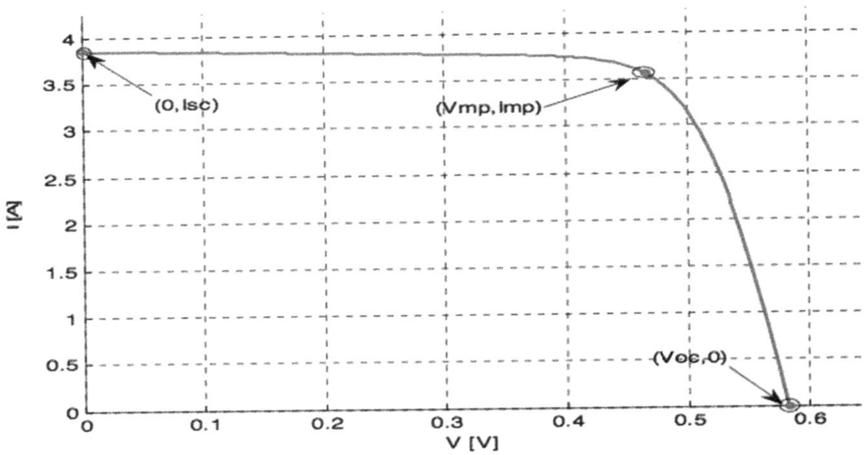

Figure 8.14. *Characteristic I(V) of a photovoltaic cell*

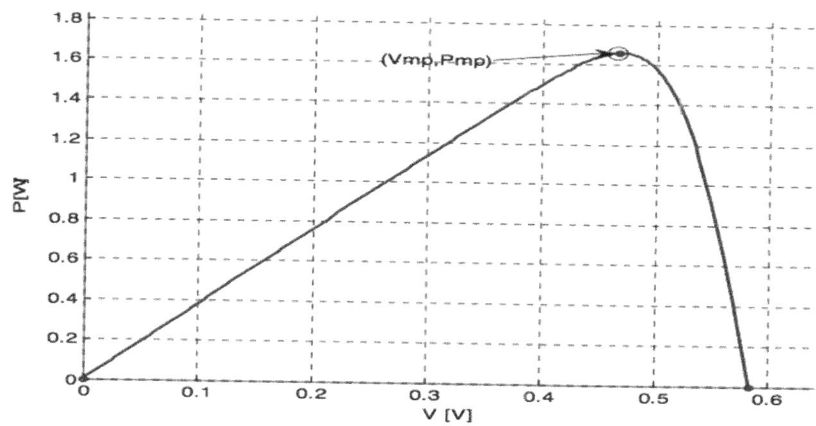

Figure 8.15. *Characteristic P(V) of a photovoltaic cell*

The power supplied by a photovoltaic panel depends on the operating point of the characteristic and is derived from the product of the voltage between its terminals and the output current. Let us note that there is a particular point where the power is at its maximum. The aim is then to place a maximum power point at this operating point in order to optimize the conversion efficiency of the photovoltaic cell.

According to the characteristic $I = f(V)$, three points are important for which the values are known: the short-circuit point, $I = I_{sc\_STC}$ and $V = 0$; the open-circuit point, $I = 0$ and $V = V_{oc\_STC}$ and the maximum power point; $I = I_{mp\_STC}$ and $V = V_{mp\_STC}$.

Each photovoltaic panel has particular characteristics, which are indicated in the data sheet supplied by the manufacturer. Important parameters that define a solar panel and allow us to calculate its performance are always specified.

The whole point of the model is to determine all unknown parameters of the extremal characteristic of a photovoltaic cell according to data specific to the cell, as provided by the manufacturer.

In what follows, we will focus on a specific photovoltaic panel model, the 30W Ameresco Solar 30J.

The calculation parameters for this type of panel, which are used for evaluating the characteristics and estimating the dynamics, are presented in Table 8.1

T_{STC}	Standard temperature under test conditions	25°C
G_{STC}	Standard irradiation under test conditions	1,000 W·m^{-2}
$V_{\mathrm{oc\_STC}}$	Open circuit voltage	22.1 V
$I_{\mathrm{sc\_STC}}$	Short circuit current	1.74 A
$P_{\mathrm{mp\_STC}}$	Maximum power	30 W
$V_{\mathrm{mp\_STC}}$	Voltage at $P_{\mathrm{mp\_STC}}$	17.9 V
$I_{\mathrm{mp\_STC}}$	Current at $P_{\mathrm{mp\_STC}}$	1.68 A
K_{i}	Temperature coefficient at $I_{\mathrm{sc\_STC}}$	0.105%/°C
K_{v}	Temperature coefficient at $V_{\mathrm{oc\_STC}}$	–0.360%/°C
N_{s}	Cells in series	36

Table 8.1. *Characteristic parameters of the photovoltaic panel*

The photovoltaic modules supplied by manufacturers represent an assembly in series and parallel of photovoltaic cells. The N_s and N_p parameters that are specified in the data sheet, respectively, represent the number of cells in series and parallel. Generally, photovoltaic modules only contain cells connected in series ($N_p = 1$).

The formula of the two-diode model characteristic is, in this case, expressed by:

$$I = I_{ph} - I_{01}\left(e^{\frac{V+IR_S}{\alpha_1 V_T N_S}} - 1\right) - I_{02}\left(e^{\frac{V+IR_S}{\alpha_2 V_T N_S}} - 1\right) - \frac{V + IR_S}{R_p}. \qquad [8.59]$$

In the following, this new expression is considered and a photovoltaic module with two diodes is modeled. Let us note that:

$$V_T = N_s \frac{k_b}{q} T. \qquad [8.60]$$

In order to determine the behavior of a photovoltaic cell, we calculate the model with two diodes, 2-D. For this, unknown parameters must be determined, in this case I_{ph}, I_{01}, I_{02}, a_1, a_2, R_s and R_p.

Let us consider the equation:

$$I_{01} = I_{02} = \frac{I_{sc\_STC} + K_i \Delta T}{e^{\frac{V_{oc\_STC} + K_v \Delta T}{\frac{\alpha_1 + \alpha_2}{p} V_T}} - 1}. \quad [8.61]$$

The short-circuit current Isc_stc is measured several times and recorded on the actual panel. The ideality factors of diodes a_1 and a_2, respectively, represent the components of diffusion and recombination of the current. The resulting experimentally deduced values are:

$$\frac{a_1 + a_2}{p} = 1 \text{ and we consider that } a_1 = 1 \quad p = 2.2. \quad [8.62]$$

We then have the possibility to evaluate the currents I_{01}, I_{02}:

$$I_{01} = I_{02} = \frac{I_{sc\_STC} + K_i \Delta T}{e^{\frac{V_{oc\_STC} + K_v \Delta T}{V_T}} - 1}. \quad [8.63]$$

The values of the resistors R_s and R_p are obtained by iteration techniques. Multiple methods have been used to evaluate these two parameters independently, but the results are not always satisfactory. In this approach, R_s and R_p are thus simultaneously calculated.

Indeed, according to the characteristic equation of the two-diode model, at the point of maximum power, R_p can be expressed as a function of R_s in the following manner:

$$R_p = \frac{V_{mp\_STC} + I_{mp\_STC} R_S}{I_{pv} - I_{01}\left(e^{\frac{V_{mp\_STC} + I_{mp\_STC} R_S}{\alpha_1 V_T}} - 1\right) - I_{02}\left(e^{\frac{V_{mp\_STC} + I_{mp\_STC} R_S}{\alpha_2 V_T}} - 1\right) - \frac{P_{mp\_E}}{V_{mp\_STC}}}. \quad [8.64]$$

The aim of this algorithm is to compare the value of the calculated maximum power point P_{mp_E} for particular values of R_S and R_p with the value of the experimental maximum power point P_{mp_E} (which exists in the manufacturer's data sheet). The idea is then to find the values of these two parameters for which the calculated maximum power is closest to the experimental maximum power.

Using the proposed mechanism, we calculate the values of the unknown parameters for the specified 2-D model, which are presented in Table 8.2.

$I_{01} = I_{02}$	I_{ph}	a_1	a_2	R_s	R_p
$7.31 \cdot 10^{-11} A$	$1.74A$	1	1.3	0.903Ω	533.059Ω

Table 8.2. *Parameters obtained for the 2-D model*

Once the unknown parameters have been determined, the characteristic equation remains to be solved in order to draw the characteristic of I as a function of voltage V, and then draw the curve that we are interested in – the curve of power P as a function of voltage V – in order to determine the maximum power voltage. We used the specified solar panel, 30W Ameresco Solar 30J, for which the required data are shown in the presented data sheet.

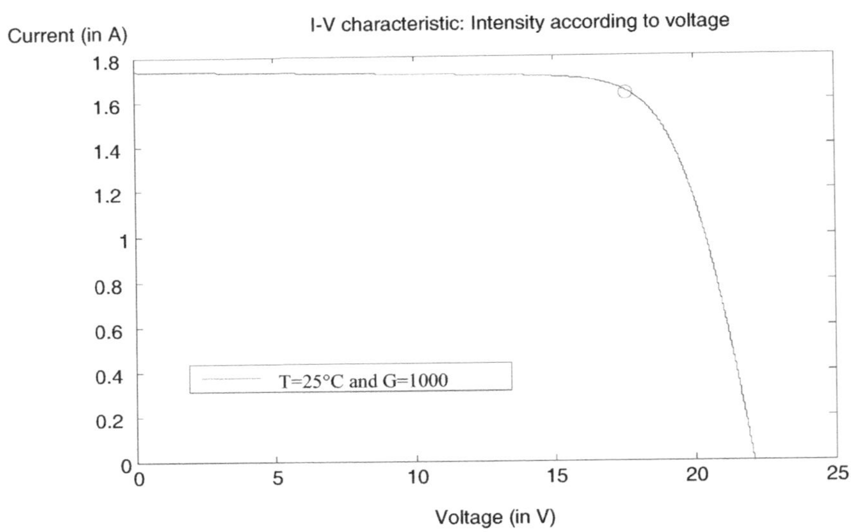

Figure 8.16. *Calculated characteristic, I(V)*

Modeling has made it possible to plot the characteristic curves as a function of parameters provided by the manufacturer, but also of parameters calculated using the formulas and methods described above, as shown in Figures 8.16 and 8.17.

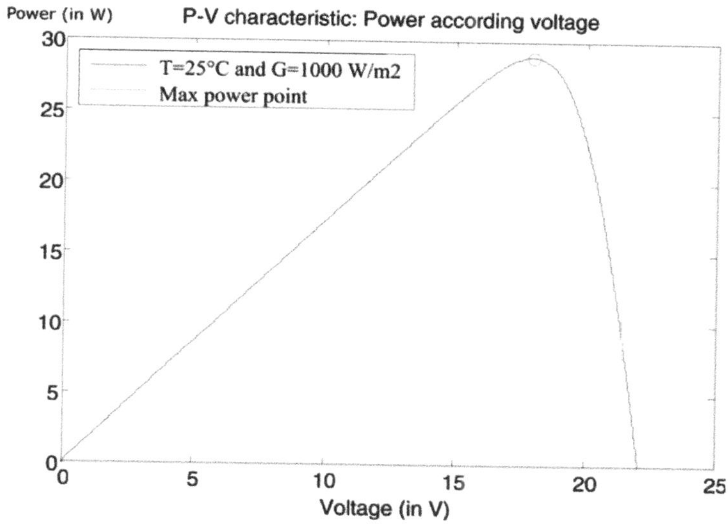

Figure 8.17. *Calculated characteristics, P(V)*

Now, let us compare the values of the calculated characteristics with those provided by the manufacturer in order to validate our approach.

Temperature T and solar radiation G have a direct impact on the performance of a photovoltaic module. When the temperature of the cells increases, the open circuit voltage decreases substantially while the short-circuit current only increases slightly. We then see that the photovoltaic cells perform better in an environment with normal temperature and clear skies, as opposed to a warmer environment.

In Figure 8.18, several curves were plotted at different temperature values, and with constant solar radiation, using the model from the previous solar panel. The solar panel power largely depends on the effects of cell radiation. Only a small fraction of the irradiation that affects the module is converted into electricity, most of the incident energy is absorbed and converted into heat.

In Figure 8.19, several curves for different radiation values were plotted using the modeling of the two-diode model at constant temperature.

Many perturbations can alter the optimal operating point of a solar panel, such as the change in temperature, solar radiation and even cloud movement. In light of this, it is of major interest to obtain the maximum point, which is the point at which the maximum amount of power can be supplied and thus the energy conversion of the photovoltaic panel can be optimized.

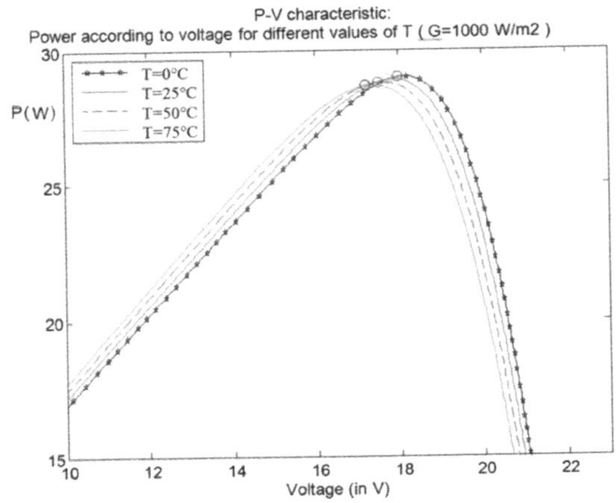

Figure 8.18. *Characteristics calculated at different temperature values with constant illumination. For a color version of this figure, see http://www.iste.co.uk/popescu/controldesign.zip*

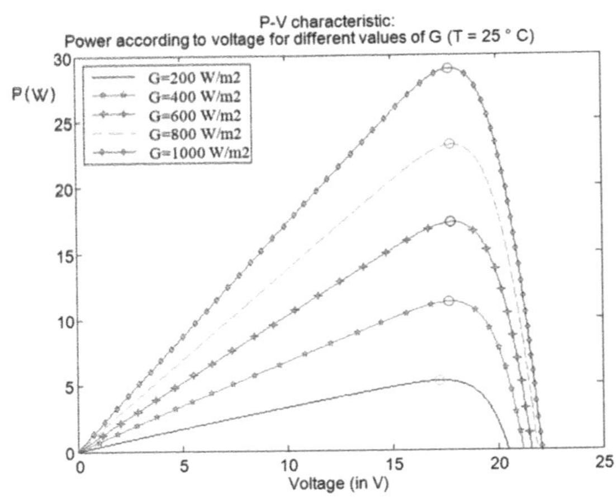

Figure 8.19. *Characteristics calculated at different values of illumination at constant temperature. For a color version of this figure, see http://www.iste.co.uk/popescu/controldesign.zip*

To calculate the maximum power point, there are many algorithms. The performance criteria of these algorithms include implementation complexity, ability of the algorithm to detect the extreme point and the rate of convergence. Two validated algorithms are proposed in our applications:

– the Coggin's algorithm, which involves approximating the power characteristic $P = f(V)$ around the optimal point through an interpolation polynomial and thus easily calculating the maximum point of this polynomial. The approximation of the maximum power point is thus achieved by the maximum of the interpolation polynomial. This operation can be repeated until the two points are very close, separated by the desired accuracy;

– the optimal gradient algorithm (Cauchy) is recognized as a robust algorithm with the fastest convergence rate. In the case of a photovoltaic panel, the optimization criterion function is the power generated as a function of two parameters: intensity and voltage, $P(I, V)$.

The algorithm's initialization parameters are as follows: a given starting point (I_0, V_0), the direction of the power gradient $P(I,V)$, the length of the search step and the search accuracy ε, which ensures the termination of the algorithm [TER 90, FIL 09, BOR 13].

8.4.1. *Extremal control of a photovoltaic panel*

The real aim of this study is to control a variable-orientation photovoltaic panel using a motor-actuator, such that it can be manipulated to follow the direction of the solar radiation, to ensure maximum output power [TOR 15, TOR 16, MIR 15, MIR 16].

In this case, the object to be controlled is the assembly, which consists of the actuator (a direct current electric motor), the photovoltaic panel and the sensor that will measure the delivered power. The purpose of the closed-loop system (Figure 8.20) is to control in real time the output power of the panel at the setpoint given by the value of the maximum power that was previously calculated.

To achieve this, the dynamic model of the assembly comprising the motor, the photovoltaic panel and the power sensor must be evaluated and the appropriate controller must be determined in order to validate the performance of the system using a simulation.

To evaluate the model of the DC motor (Figure 8.21), two equations are used.

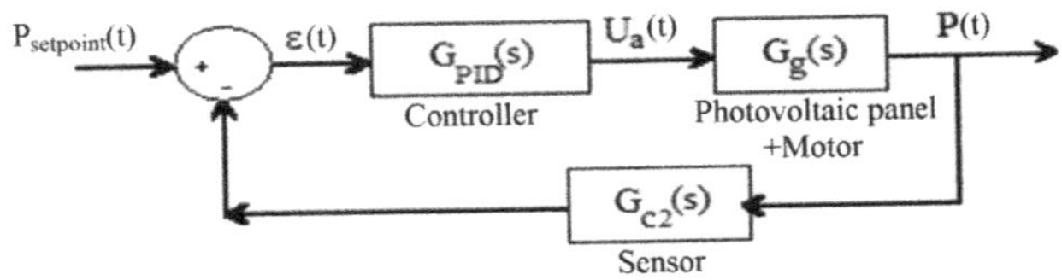

Figure 8.20. *Digital system for maximum power control*

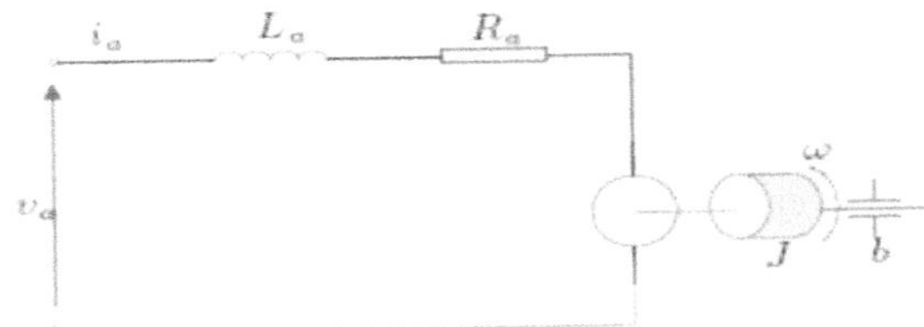

Figure 8.21. *Electromechanical diagram of a DC motor*

We obtain the electric equation:

$$u_a(t) = u_{Ra}(t) + u_{La}(t) + e_a(t) = R_a i_a(t) + L_a \frac{di_a(t)}{dt} + e_a(t) \quad [8.65]$$

and the mechanical equation

$$J\frac{d\omega(t)}{dt} = C_m(t) - C_f(t). \quad [8.66]$$

For the chosen motor, we obtain the following transfer function:

$$G_m(s) = \frac{\Omega(s)}{U_a(s)} = \frac{\frac{K}{L_d J}}{s^2 + \frac{R_a J + L_a f}{L_a J}s + \frac{R_a f + K^2}{L_a J}} \quad [8.67]$$

and after a few calculations based on the parameters of the motor, we obtained:

$$G_m(s) = \frac{13.44}{s^2 + 93.97s + 40.33} = \frac{K_m}{(1+T_{m1}s)(1+T_{m2}s)} = \frac{0.31}{(1+0.01s)(1+2.17s)}. \quad [8.68]$$

Experimental simulations show that the power delivered by the photovoltaic panel is proportional to the solar radiation received. For the dynamics of the selected panel, based on its functional characteristics, the following normalized transfer function (gain) is obtained:

$$G_{pv}(s) = K_{pv} = 1.03\,. \qquad [8.69]$$

The angle of incidence α_i, which corresponds to the plane formed between the solar panel and the rays of light, is very important. The optimal angle of incidence corresponds to a value of 90°. Each time this angle is modified, the surface of the solar panel that is exposed to the light rays decreases, thus the radiation received decreases. The coefficient of radiation of a solar panel as a function of the inclination from the rays is expressed by the sine of the angle of incidence. In our case, the role of the motor is to orient the panel to follow the incident rays. Thus, angle α_i is very small and we can deduce that the radiation received by the panel is proportional to the angle of incidence α_i. The transfer function between the angle of incidence and the angular position of the motor is obviously an integrator type and thus, we obtain:

$$\frac{\alpha_i}{\Omega} = \frac{1}{s}\,. \qquad [8.70]$$

A sensor is used to measure the power delivered by the photovoltaic panel, which at the output gives a standard signal that is proportional to the power received. The proportional behavior of the sensor is expressed by the calculated gain:

$$G_c(s) = K_c = 1.15\,. \qquad [8.71]$$

The overall transfer function of the motor–panel–sensor assembly, in continuous regime, gives the following expression:

$$G_g(s) = \frac{0.278}{s(1+0.011s)(1+2.17s)}\,. \qquad [8.72]$$

The aim of the extremal system is to ensure the value of the power generated by the panel at a calculated setpoint, which corresponds to the maximum power point. To do this, we use the previous results to evaluate the maximum power point required to design the servo system. The general configuration of the system is shown in Figure 8.22.

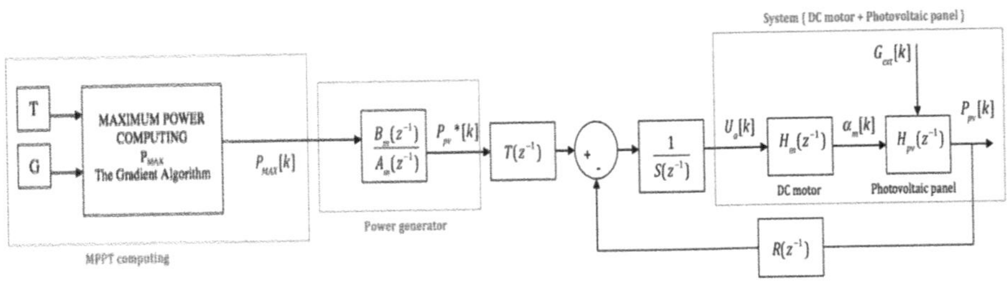

Figure 8.22. *General diagram of the extremal system*

By discretizing the transfer function [8.72] with a sampling period of 0.05 s, we obtain:

$$G(z^{-1}) = \frac{1.364 \cdot 10^{-5} z^{-3} + 2.201 \cdot 10^{-4} z^{-2} + 1.183 \cdot 10^{-4} z^{-1}}{9.111 \cdot 10^{-3} z^{-3} + 0.9971 \cdot z^{-2} - 1.988 z^{-1} + 1}. \qquad [8.73]$$

The performance of the control system is specified by the desired closed-loop poles that are introduced using the characteristic polynomial $P(z^{-1})$, which represents the discrete equivalent of the continuous model specified by the set ($\zeta = 0.87,\ \omega_0 = 5$ rad/s).

Using the dedicated WINREG software, the control polynomials R and S are easily calculated, based on the characteristic polynomial $P(z^{-1})$:

$$P(z^{-1}) = (1 - 1.596847z^{-1} + 0.647265z^{-2})(1 - 1.058z^{-1} + 0.81z^{-2}). \qquad [8.74]$$

We then obtain:

$$\begin{aligned} R(z^{-1}) &= 1193.887 - 1106.043z^{-1} - 10.065z^{-2} \\ S(z^{-1}) &= 1 + 0.2499z^{-1} + 0.0150z^{-2} \end{aligned}. \qquad [8.75]$$

Additional zeros are introduced in S by the polynomial $G_S(z^{-1}) = (1 - z^{-1})(1 - 1.756z^{-1} + z^{-2})$ and in R, by the polynomial $G_R(z^{-1}) = (1 + z^{-1})$, to ensure the robustness of the control algorithm.

The generated trajectory is expressed through a model that receives the value of the maximum power as input and which has the desired power dynamics at its output. This is a second-order model defined by the following parameters: $\zeta = 0.8$ and $\omega_0 = 3 rad.s^{-1}$.

For a sampling period of 0.05 s, the transfer function of the generator was calculated.

The precompensator T, which ensures that the tracking dynamic was evaluated on the basis of the following equations, is written as:

$$T(z^{-1}) = G \cdot P(z^{-1})$$
$$G = \begin{cases} \dfrac{1}{B(1)}; B(1) \neq 0 \\ 1; B(1) = 0 \end{cases} \qquad [8.76]$$

using the characteristic polynomial from equation [8.74].

The robust control was implemented on a microcontroller connected to a virtual process (dynamic model) in a *hardware in the loop* configuration (Arduino numerical platform). The performances of the simulated closed-loop system are shown in Figure 8.23.

The delay in the figure at the beginning of the simulation is generated by the initialization of the microcontroller.

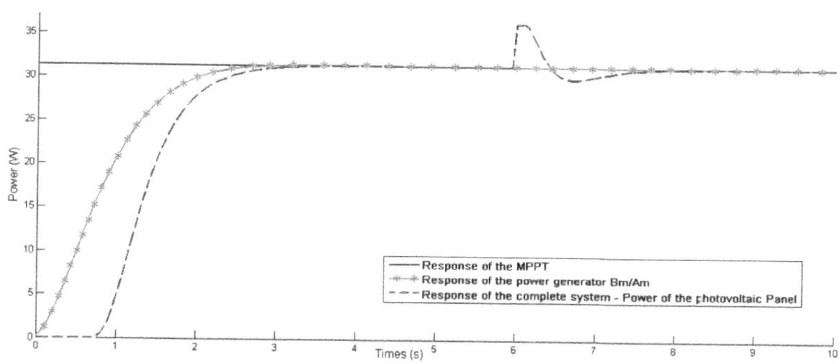

Figure 8.23. *Response of a closed-loop extremal system implemented on a microcontroller. For a color version of this figure, see http://www.iste.co.uk/popescu/controldesign.zip*

The proposed solution was installed in practice on an experimental photovoltaic platform integrated in an intelligent local network [MIR 15, MIR 16, TOR 15, TOR 16].

Appendix A

Matrix Transformation from Any Representation to the Companion Form or Arrow Form

Let us consider the process for which the evolution is described by equation:

$$\dot{x} = Ax + B_u \qquad [A.1]$$

with the matrix of change of basis P_c such that

$$x = P_c x_c \qquad [A.2]$$

makes it possible to obtain a representation of the system described by equation (S') in controllable form, with x_c representing the state vector of the process in the new base. We obtain:

$$\dot{x}_c = A_c x_c + B_c u \qquad [A.3]$$

$$A_c = P_c^{-1} A_l P_c, \quad B_c = P_c^{-1} B_l \,. \qquad [A.4]$$

Let us note $\mathbb{P}_A(\lambda)$, the characteristic polynomial of matrix A_c:

$$\mathbb{P}_A(\lambda) = a_0 + a_1\lambda + \cdots + a_{n-1}\lambda^{n-1} + \lambda^n \,. \qquad [A.5]$$

If the pair $(A(x_0), B(x_0))$ is controllable, the search for a description of the system in controllable form allows us to characterize the system through the pair (A_c, B_c), which is defined by:

$$A_c = \begin{bmatrix} 0 & 1 & 0 & \cdots & \cdots & \cdots & 0 \\ 0 & 0 & 1 & \ddots & & & 0 \\ \vdots & \vdots & \vdots & \ddots & \ddots & \vdots & \vdots \\ \vdots & \vdots & \vdots & \ddots & \ddots & \ddots & \vdots \\ 0 & 0 & 0 & \cdots & \cdots & \ddots & 0 \\ 0 & 0 & 0 & \cdots & \cdots & \cdots & 1 \\ -a_0 & -a_1 & -a_2 & \cdots & \cdots & \cdots & -a_{n-1} \end{bmatrix} B_c = \begin{bmatrix} 0 \\ 0 \\ \vdots \\ \vdots \\ 0 \\ 0 \\ 1 \end{bmatrix}. \quad \text{[A.6]}$$

We obtain:

$$P_c A_c = A P_c, \quad P_c B_c = B. \quad \text{[A.7]}$$

Let us note P_{ci}, the ith column of P_c such that:

$$P_c = [P_{c1} \; P_{c2} \ldots P_{cn}]. \quad \text{[A.8]}$$

The P_c columns are deduced from the following algorithm:

$$\begin{aligned} P_{c_n} &= B \\ P_{c_{n-1}} &= (A + a_{n-1} I) B \\ P_{c_{n-2}} &= (A^2 + a_{n-1} A + a_{n-2} I) B, \\ &\vdots \\ P_{c_1} &= (A^{n-1} + a_{n-1} A^{n-2} + \cdots + a_1 I) B, \end{aligned} \quad \text{[A.9]}$$

we obtain:

$$A_{cf} = P_{cf}^{-1} A_c P_{cf} \quad \text{[A.10]}$$

$$B_{cf} = P_{cf}^{-1} B_c \quad \text{[A.11]}$$

A1.1. Transition from a companion matrix to an arrow form matrix

Let us note P_{cf}, which is the transition matrix from the companion form to the arrow form.

We obtain:

$$A_{cf} = P_{cf}^{-1} A_c P_{cf} \qquad \text{[A.12]}$$

$$P_{cf} = \begin{bmatrix} 1 & 1 & \cdots & 1 & 0 \\ \alpha_1 & \alpha_2 & \cdots & \alpha_{n-1} & 0 \\ \vdots & \vdots & \vdots & \vdots & \vdots \\ \alpha_1^{n-2} & \alpha_2^{n-2} & \cdots & \alpha_{n-1}^{n-2} & 0 \\ \alpha_1^{n-2} & \alpha_2^{n-1} & \cdots & \alpha_{n-1}^{n-1} & 1 \end{bmatrix}, A_{cf}(.) = \begin{bmatrix} \alpha_1 & & \cdots & & \beta_1 \\ \vdots & \alpha_2 & & & \beta_2 \\ \vdots & \vdots & \ddots & & \vdots \\ \vdots & \vdots & & \alpha_{n-1} & \beta_{n-1} \\ \gamma_{i,1} & \gamma_{i,2} & \cdots & \gamma_{i,n-1} & \gamma_{i,n} \end{bmatrix} \qquad \text{[A.13]}$$

with

$$\begin{aligned} \beta_j &= \prod_{k=1}^{n-1} (\alpha_j - \alpha_k)^{-1}; \quad \forall j = 1, 2 \ldots n-1 \\ \gamma_{i,n} &= -a_{n-1} - \sum_{i=1}^{n-1} \alpha_i \\ \gamma_{i,j} &= -\mathbb{P}(\alpha_j); \quad \forall j = 1, 2 \ldots n-1 \\ \mathbb{P}(\alpha_j) &= a_0 + a_1\alpha_i + \cdots + a_{n-1}\alpha^{n_i-1} + \alpha^n \end{aligned} \qquad . \qquad \text{[A.14]}$$

The transition of a matrix of any shape to the arrow form can be achieved by successively performing two basis changes:

$$x = P_f x_f \, . \qquad \text{[A.15]}$$

We obtain:

$$A_f = P_f^{-1} A(x) P_f \qquad \text{[A.16]}$$

and

$$B_f = P_f^{-1} B \qquad \text{[A.17]}$$

with

$$P_f^{-1} = \left(P_c P_{cf}\right)^{-1} \text{ and } P_f = P_c P_{cf} . \qquad [A.18]$$

A1.2. Direct transition of a matrix of any form to an arrow form

Let A and B be a controllable pair $x \in \mathbb{R}^n, u \in \mathbb{R}, A \in \mathbb{R}^{n \times n}, B \in \mathbb{R}^n$, and let $\mathbb{P}(.,\lambda)$ be the characteristic polynomial of A:

$$\mathbb{P}(\lambda) = \lambda^n + \sum_{i=0}^{n-1} a_i \lambda^i . \qquad [A.19]$$

The change of basis defined by matrix P_1 is

$$P_1 = [P_1^1 \quad P_1^2 \quad \dots \quad P_1^n] \qquad [A.20]$$

such that

$$x = P_1 y \qquad [A.21]$$

with

$$\begin{cases} P_1^{\mathrm{i}} = \left[A^{n-1} + \sum_{k=1}^{n-1} \alpha_{k,i} A^{k-1} \right] B & i = 1,\dots,n-1 \\ P_1^{\mathrm{n}} = PB \end{cases} \qquad [A.22]$$

$$\alpha_{k,i} = a_k + \left(f_{i,i}\right)^{n-k} + \sum_{j=1}^{n-k-1} a_{k+j} \left(f_{i,i}\right)^j, \quad k = 1,\dots,n-1, i = 1,\dots,n-1, \qquad [A.23]$$

where $f_{i,i}, \forall i = 1,\dots,n-1$, are distinct non-zero arbitrary parameters, making it possible to describe matrix A in an arrow form $\mathrm{F} = P_1^{-1} A P_1$.

$$\mathrm{F} = \begin{bmatrix} f_{11} & & & f_{1N} \\ & \ddots & & \vdots \\ & & f_{(k-1)(k-1)} & f_{(N-1)n} \\ f_{n1} & \dots & f_{n(N-1)} & f_{nn} \end{bmatrix} \qquad [A.24]$$

with

$$
\begin{aligned}
f_{in} &= \prod_{j=1}^{n-1} (f_{ii} - f_{jj})^{-1} && \forall i = 1, 2, \ldots n-1 \\
f_{ni} &= -\mathbb{P}(., f_{ii}) && \forall i = 1, 2, \ldots n-1\,. \\
f_{nn} &= -a_{n-1}(.) - \sum_{i=1}^{n-1} f_{ii}
\end{aligned}
\qquad [A.25]
$$

Appendix B

Determination of the Maximum Error for Pole Placement for a Nonlinear Third-Order Process

Let us consider the process (S) for which the evolution is described by equation:

$$(S):\quad \dot{x}_c = Ax_c + B_c u \qquad [B.1]$$

with

$$\det(\lambda I - A(0)) = \det\begin{bmatrix} \lambda+10.5 & 0 & 0 \\ 0 & \lambda+3.4 & -3.6 \\ 0 & 0.6 & \lambda+7.6 \end{bmatrix} = \lambda^3 + 21.5\lambda^2 + 143.5\lambda + 294 \qquad [B.2]$$

$$B(x(t)) = \begin{bmatrix} 0.5 + 0.01\sin x_1 \\ 0.4 \\ 0.02\,\mathrm{sat}\,x_1 + 0.3 \end{bmatrix} \qquad [B.3]$$

and

$$|B| \leq B_m \qquad [B.4]$$

with

$$B_m = \begin{bmatrix} 0.2 \\ 0.15 \\ 0.1 \end{bmatrix}. \qquad [B.5]$$

By linearizing the system without uncertainties around the operating point ($x_0 = 0$), we obtain:

$$\dot{y}(t) = A(x_0)y + B(x_0)u \quad \text{[B.6]}$$

with

$$A(0) = \begin{bmatrix} -10.5 & 0 & 0 \\ 0 & -3.4 & 3.6 \\ 0 & -0.6 & -7.6 \end{bmatrix}; B(0) = \begin{bmatrix} 0,5 \\ 0,4 \\ 0,3 \end{bmatrix} \quad \text{[B.7]}$$

The characteristic polynomial of the system is:

$$\det(\lambda I - A(0)) = \det\begin{bmatrix} -10.5 & 0 & 0 \\ 0 & -3.4 & 3.6 \\ 0 & -0.6 & -7.6 \end{bmatrix} = \lambda^3 + 21.5\lambda^2 + 143.5\lambda + 294 \quad \text{[B.8]}$$

such that:

$$P_A(\lambda) = \lambda^3 + 21.5\lambda^2 + 143.5\lambda + 294 \quad \text{[B.9]}$$

With the base change defined by matrix P_c:

$$P_c = \begin{bmatrix} 14 & 5.5 & 0.5 \\ 43.26 & 8.32 & 0.4 \\ 8.19 & 3.93 & 0.3 \end{bmatrix} \text{ and } P_c^{-1} = \begin{bmatrix} 0.0879 & 0.03 & -0.1865 \\ -0.9231 & 0.01 & 1.5251 \\ 9.6923 & -0.9491 & -11.551 \end{bmatrix} \quad \text{[B.10]}$$

by letting

$$A_c = P_c^{-1}A(0)P_c \quad \text{[B.11]}$$

$$B_c = P_c^{-1}B(0) \quad \text{[B.12]}$$

we obtain a canonical controllable form

$$\dot{x}_c = A_c x_c + B_c u \quad \text{[B.13]}$$

with the command

$$u = -L_c x \quad \text{[B.14]}$$

$$A_c = \begin{bmatrix} 0 & 1 & 0 \\ 0 & 0 & 1 \\ -294 & -143.5 & -21.5 \end{bmatrix}; \quad B_c = \begin{bmatrix} 0 \\ 0 \\ 1 \end{bmatrix}. \quad \text{[B.15]}$$

To determine the feedback state gain L, let us consider the closed-loop system described in the state space by:

$$\dot{x}_c(t) = (A_c - B_c L_c) x_c(t) \quad \text{[B.16]}$$

$$L_c = \begin{bmatrix} l_{c_0} & l_{c_1} & l_{c_2} \end{bmatrix} \quad \text{[B.17]}$$

$$\dot{x}_c = \left(\begin{bmatrix} 0 & 1 & 0 \\ 0 & 0 & 1 \\ -294 & -143.5 & -21.5 \end{bmatrix} - \begin{bmatrix} 0 \\ 0 \\ 1 \end{bmatrix} \begin{bmatrix} l_{c_0} & l_{c_1} & l_{c_2} \end{bmatrix} \right) x_c . \quad \text{[B.18]}$$

Let us choose the poles corresponding to the polynomial characteristic $\mathrm{P}_{A_c - B_c L_c}(\lambda)$ of the form:

$$\mathrm{P}_{A_c - B_c L_c}(\lambda) = (\lambda + 8)(\lambda + 9)(\lambda + 10) = \lambda^3 + 27\lambda^2 + 242\lambda + 720 . \quad \text{[B.19]}$$

We obtain

$$\begin{aligned} \dot{x}_c &= (A_c - B_c L_c) x_c = \mathcal{A}_c x_c \\ &= \left(\begin{bmatrix} 0 & 1 & 0 \\ 0 & 0 & 1 \\ -294 & -143.5 & -21.5 \end{bmatrix} - \begin{bmatrix} 0 \\ 0 \\ 1 \end{bmatrix} \begin{bmatrix} l_{c_0} & l_{c_1} & l_{c_2} \end{bmatrix} \right) x_c \quad \text{[B.20]} \\ &= \begin{bmatrix} 0 & 1 & 0 \\ 0 & 0 & 1 \\ -720 & -242 & -27 \end{bmatrix} x_c \end{aligned}$$

hence

$$u = -\begin{bmatrix} l_{c_1} & l_{c_2} & l_{c_3} \end{bmatrix} x_c \quad \text{[B.21]}$$

with

$$L_c = \begin{bmatrix} 426 & 98.5 & 5.5 \end{bmatrix} \quad \text{[B.22]}$$

such that

$$L = L_c P_c^{-1} = \begin{bmatrix} -0.1648 & 8.5315 & 7.2328 \end{bmatrix}. \quad \text{[B.23]}$$

Let

$$\mathcal{A}(x) = (A(x) - B(x)L) \;. \quad \text{[B.24]}$$

Linearization of the closed-loop system around the operating point x=0

Let $\mathcal{A}(0)$ be the matrix of the linearized closed-loop system without input:

$$\begin{aligned} \mathcal{A}(0) &= A(0) - B(0)L \\ &= \begin{bmatrix} -10.5 & 0 & 0 \\ 0 & -3.4 & 3.6 \\ 0 & -0.6 & -7.6 \end{bmatrix} - \begin{bmatrix} 0.5 \\ 0.4 \\ 0.3 \end{bmatrix} \begin{bmatrix} -0.1648 & 8.5315 & 7.2328 \end{bmatrix} \quad \text{[B.25]} \\ &= \begin{bmatrix} -10.4176 & -4.2657 & -3.6164 \\ 0.0659 & -6.8126 & 0.7069 \\ 0.0495 & -3.1594 & -9.7698 \end{bmatrix} \end{aligned}$$

Calculation of the base change that diagonalizes $\mathcal{A}(0)$

Let P be this base change. Thus, we obtain:

$$P = P_c P_D = \begin{bmatrix} 14 & 5.5 & 0.5 \\ 43.26 & 8.32 & 0.4 \\ 8.19 & 3.93 & 0.3 \end{bmatrix} \begin{bmatrix} 1 & 1 & 1 \\ -8 & -9 & -10 \\ 64 & 81 & 100 \end{bmatrix} = \begin{bmatrix} 2 & 5 & 9 \\ 2.3 & 0.78 & 0.06 \\ -4.05 & -2.88 & -1.11 \end{bmatrix} \quad \text{[B.26]}$$

hence

$$P^{-1} = \begin{bmatrix} 0.033 & 0.969 & 0.3197 \\ -0.1099 & -1.6284 & -0.9790 \\ 0.1648 & 0.6893 & 0.4729 \end{bmatrix}. \quad [B.27]$$

We obtain

$$\mathcal{A}_D(x) = P^{-1}\mathcal{A}(x)\ P; \quad B_D = P^{-1}B' \quad [B.28]$$

thus

$$\dot{x}_D = \mathcal{A}_D(x)x_D + B_D \quad [B.29]$$

$$\mathcal{A}(x) = (A(x) - B(x)L) \quad [B.30]$$

with

$$\mathcal{A}(x(t)) = \begin{bmatrix} a_{11} & a_{12} & a_{13} \\ a_{21} & a_{22} & a_{23} \\ a_{31} & a_{32} & a_{33} \end{bmatrix} \quad [B.31]$$

$$a_{11} = 0.0016\sin x_1 - 0.5\cos x_1 - 9.9176$$
$$a_{12} = -0.0853\sin x_1 - 4.2657$$
$$a_{13} = 0.04(\cos x_1 - e^{-x_2^2}) - 0.0723\sin x_1 - 3.6164$$
$$a_{21} = -0.01\frac{x_1 x_2}{x_1^2 x_2^2}\sin x_1 + 0.0659$$
$$a_{22} = 0.12\cos x_1 - 6.9326$$
$$a_{23} = 0.32e^{-x_1^2} + 0.3869$$
$$a_{31} = 0.0033\text{sat}x_1 + 0.01\sin x_1 + 0.0495$$
$$a_{32} = -0.1706\text{sat}x_1 - 0.6e^{-x_2^2} - 2.5594$$
$$a_{33} = -0.1447\text{sat}x_1 - 0.22\cos x_1 - 9.5498$$

we obtain

$$\mathcal{A}_D(x) = P^{-1}\mathcal{A}(x)\ P \quad [B.32]$$

$$\mathcal{A}_D(x(t)) = \begin{bmatrix} a_{d11} & a_{d12} & a_{d13} \\ a_{d21} & a_{d22} & a_{d23} \\ a_{d31} & a_{d32} & a_{d33} \end{bmatrix} \quad [B.33]$$

with

$$a_{d11} = 0.0639\text{sat}x_1 - 1.697e^{-x_1^2} + 0.0053e^{-x_2^2} + 0.514\cos x_1 + 0.0033\sin x_1 + 0.0064\sin x_3 - 0.0194\frac{x_1x_2}{x_1^2+x_2^2}\sin x_1 - 6.8223$$

$$a_{d12} = 0.0959\text{sat}x_1 - 1.0427e^{-x_1^2} + 0.0038e^{-x_2^2} + 0.207\cos x_1 + 0.0049\sin x_1 + 0.016\sin x_3 - 0.0485\frac{x_1x_2}{x_1^2+x_2^2}\sin x_1 + 0.8318$$

$$a_{d13} = 0.0575\text{sat}x_1 - 0.35571e^{-x_1^2} + 0.0015e^{-x_2^2} - 0.0648\cos x_1 + 0.003\sin x_1 + 0.0288\sin x_3 - 0.0872\frac{x_1x_2}{x_1^2+x_2^2}\sin x_1 + 0.419$$

$$a_{d21} = -0.1958\text{sat}x_1 + 3.4614e^{-x_1^2} - 0.0178e^{-x_2^2} - 1.194\cos x_1 - 0.011\sin x_1 - 0.0196\sin x_3 + 0.0326\frac{x_1x_2}{x_1^2+x_2^2}\sin x_1 - 2.2496$$

$$a_{d22} = -0.2937\text{sat}x_1 + 1.9589e^{-x_1^2} - 0.0127e^{-x_2^2} - 0.4853\cos x_1 - 0.0165\sin x_1 - 0.049\sin x_3 + 0.0814\frac{x_1x_2}{x_1^2+x_2^2}\sin x_1 - 10.4609$$

$$a_{d23} = -0.1762\text{sat}x_1 + 0.6136e^{-x_1^2} - 0.0049e^{-x_2^2} + 0.2486\cos x_1 - 0.0099\sin x_1 - 0.0881\sin x_3 + 0.1466\frac{x_1x_2}{x_1^2+x_2^2}\sin x - 0.8573$$

$$a_{d31} = 0.0946\text{sat}x_1 - 1.5459e^{-x_1^2} + 0.0267e^{-x_2^2} + 0.42\cos x_1 + 0.0165\sin x_1 + 0.095\sin x_3 - 0.0138\frac{x_1x_2}{x_1^2+x_2^2}\sin x_1 + 1.0992$$

$$a_{d32} = 0.1419\text{sat}x_1 - 0.8566e^{-x_1^2} + 0.019e^{-x_2^2} + 0.067\cos x_1 + 0.0247\sin x_1 + 0.0236\sin x_3 - 0.0275\text{sat}x_1\cos t - 0.0345\frac{x_1x_2}{x_1^2+x_2^2}\sin x_1 + 0.9045$$

$$a_{d33} = 0.0851\text{sat}x_1 - 0.2619e^{-x_1^2} + 0.0073e^{-x_2^2} - 0.6286\cos x_1 + 0.0148\sin x_1 + 0.0426\sin x_3 - 0.062\frac{x_1x_2}{x_1^2+x_2^2}\sin x_1 - 9.1168$$

By overvaluing $M(.)$ by M and $N(.)$ by N, we obtain the comparison system:

$$z = Mz + N \quad [B.34]$$

$$N(.) = \left|P^{-1}B'(.)\right| = \left|B'_D(.)\right| \qquad [B.35]$$

$$P^{-1}B' \leq P^{-1}B'_m \qquad [B.36]$$

$$N = \max\left|P^{-1}B'(.)\right|' \leq \begin{bmatrix} 0.033 & 0.969 & 0.3197 \\ 0.1099 & 1.6284 & 0.9790 \\ 0.1648 & 0.6893 & 0.4729 \end{bmatrix} \begin{bmatrix} 0.2 \\ 0.15 \\ 0.1 \end{bmatrix} = \begin{bmatrix} 0.1839 \\ 0.3641 \\ 0.1836 \end{bmatrix} \qquad [B.37]$$

$$\dot{z}(t) = \begin{bmatrix} -6.2197 & 1.1837 & 0.6182 \\ 3.7041 & -7.6168 & 1.4583 \\ 1.7589 & 1.1973 & -8.3074 \end{bmatrix} z(t) + N \begin{bmatrix} 0.1839 \\ 0.3641 \\ 0.1836 \end{bmatrix}. \qquad [B.38]$$

With the following conditions

$$\begin{cases} -6.2197 \prec 0 \\ (-6.2197 \times -7.6168) - (3.7041 \times 1.1837) \succ 0 \\ (-1)^3 \det(M) \prec 0 \Rightarrow (-1)^3 \times (-332.2047) \succ 0 \end{cases} \qquad [B.39]$$

having been verified that M is the opposite of an M-matrix and we obtain:

$$\lim_{t \to +\infty} z(t) = -M^{-1}N \qquad [B.40]$$

$$\lim_{t \to +\infty} p(x_D(t)) \leq -M^{-1}N\,. \qquad [B.41]$$

Hence, the attractor defined by:

$$p(x_D(t)) \leq -M^{-1}N \qquad [B.42]$$

$$p(x_D(t)) \leq \begin{bmatrix} -0.1852 & -0.0318 & -0.0194 \\ -0.1003 & -0.1523 & -0.0342 \\ -0.0537 & -0.0287 & -0.1294 \end{bmatrix} \begin{bmatrix} 0.1839 \\ 0.3641 \\ 0.1836 \end{bmatrix} \qquad [B.43]$$

$$p(P^{-1}x(t)) \leq \begin{bmatrix} 0.0492 \\ 0.0802 \\ 0.0441 \end{bmatrix} \qquad [B.44]$$

Knowing that

$$\lim_{t \to \infty} p(P^{-1}x(t)) \le -M^{-1}N \qquad [B.45]$$

domain D is defined by equations:

$$\begin{cases} |0.033x_1 + 0.969x_2 + 0.3197x_3| \le 0.0492 \\ |-0.1099x_1 - 1.6284x_2 - 0.9790x_3| \le 0.0802 \\ |0.1648x_1 + 0.6893x_2 + 0.4729x_3| \le 0.0441 \end{cases} \qquad [B.46]$$

Figure B.1 shows the evolution of the system toward the calculated attractor.

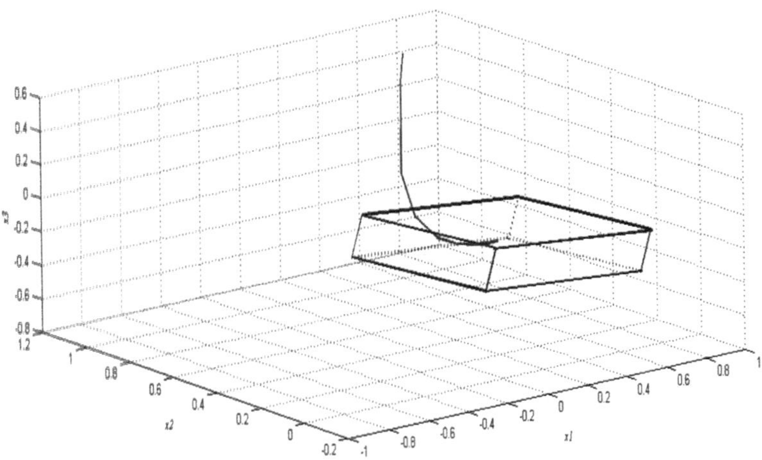

Figure B.1. *Evolution of the system toward the attraction zone*

Appendix C

Determining the Attractor in a Nonlinear Process Controlled by Linear Decoupling

Let us consider the following second-order system:

$$(S): \begin{cases} \dot{x}_1 = -x_1 + x_2{}^2 u_1 - u_2 \\ \dot{x}_2 = 2x_1 x_2 + u_1 + u_2 \\ y_1 = x_1 \\ y_2 = x_2 \end{cases} \qquad [C.1]$$

which can be described by

$$\begin{aligned} \dot{x} &= f(x) + G(x)u + \delta f \\ y &= h(x) \end{aligned} \qquad [C.2]$$

with

$$f(x) = \begin{bmatrix} -x_1 \\ 2x_1 x_2 \end{bmatrix}; \quad G(x) = \begin{bmatrix} x_2^2 & -1 \\ 1 & 1 \end{bmatrix}; \delta f = \begin{bmatrix} \delta f_1 \\ \delta f_2 \end{bmatrix}. \qquad [C.3]$$

To check whether the system without perturbation can have an input–output linearization, we need to determine the relative degree of the system by deriving y:

$$\begin{cases} \dot{y}_1 = \dot{x}_1 = -x_1 + x_2{}^2 u_1 - u_2 \\ \dot{y}_2 = \dot{x}_2 = 2x_1 x_2 + u_1 + u_2 \end{cases}. \qquad [C.4]$$

Since the input appears at the first derivative of y, $h_{i_x}^T G \neq 0$, we can stop the calculation. We obtain:

$$h^*(x) = f(x) = \begin{bmatrix} -x_1 \\ 2x_1x_2 \end{bmatrix} \qquad [C.5]$$

knowing that

$$G^* = \begin{bmatrix} x_2^2 & -1 \\ 1 & 1 \end{bmatrix} \qquad [C.6]$$

we obtain

$$G^{*^{-1}} = \frac{1}{x_2^2+1}\begin{bmatrix} 1 & 1 \\ -1 & x_2^2 \end{bmatrix}. \qquad [C.7]$$

Since G^* is inversible, we can set the input u to equal:

$$u = G^{*-1}(-h^* + v) \qquad [C.8]$$

hence

$$y^* = h^* + G^* u = v = \begin{bmatrix} v_1 \\ v_2 \end{bmatrix} \qquad [C.9]$$

we obtain:

$$\begin{bmatrix} u_1 \\ u_2 \end{bmatrix} = \frac{1}{x_2^2+1}\begin{bmatrix} 1 & 1 \\ -1 & x_2^2 \end{bmatrix}\begin{bmatrix} x_1 + v_1 \\ -2x_1x_2 + v_2 \end{bmatrix} \qquad [C.10]$$

$$\begin{bmatrix} u_1 \\ u_2 \end{bmatrix} = \frac{1}{x_2^2+1}\begin{bmatrix} x_1 + v_1 - 2x_1x_2 + v_2 \\ -x_1 - v_1 - 2x_1x_2^3 + x_2^2 v_2 \end{bmatrix}. \qquad [C.11]$$

The system's input–output behavior is thus linearized.

Control of the process leads to

$$v_i = \frac{1}{\tau_i}\left(y_i^c - y_i\right). \qquad [C.12]$$

Let $y_i = x_i$ and $y_i^c = 0$

$$\dot{x}_i = \frac{1}{\tau_i} x_i . \tag{C.13}$$

Taking the perturbations and/or uncertainties on f into account, we obtain:

$$\begin{aligned} \dot{x}_1 &= v_1 + \delta f_1 = -\frac{x_1}{\tau_1} + \delta f_1 \\ \dot{x}_2 &= v_2 + \delta f_2 = -\frac{x_2}{\tau_2} + \delta f_2 \end{aligned} \tag{C.14}$$

$$\dot{x} = \begin{bmatrix} -\frac{1}{\tau_1} & 0 \\ 0 & -\frac{1}{\tau_2} \end{bmatrix} x + \begin{bmatrix} \delta f_1 \\ \delta f_2 \end{bmatrix} \tag{C.15}$$

with

$$|\delta f_i| \leq \delta f_{N_i} . \tag{C.16}$$

For the vector norm $p(x) = \left[|x_1|, |x_2|\right]^T$, we get the following comparison system:

$$z \in \mathbb{R}^n \,/\, \dot{z} = Mz + N . \tag{C.17}$$

In this example, $M = A$ and vector N has positive coefficients with:

$$N = \delta f_N \tag{C.18}$$

we then obtain

$$\lim_{t \to \infty} |x| \leq -M^{-1} N = \begin{bmatrix} \tau_1 \delta f_{N_1} \\ \tau_2 \delta f_{N_2} \end{bmatrix} . \tag{C.19}$$

For the following data:

$$\tau_1 = \tau_2 = 1s \text{, et } \delta f_{N_1} = \delta f_{N_2} = 0.01 \tag{C.20}$$

we obtain

$$\lim_{t\to\infty} |x| \leq -M^{-1}N = \begin{bmatrix} 0.01 \\ 0.01 \end{bmatrix}. \qquad [C.21]$$

The convergence of the state variables of the considered system toward the calculated attractor is illustrated in Figure C.1.

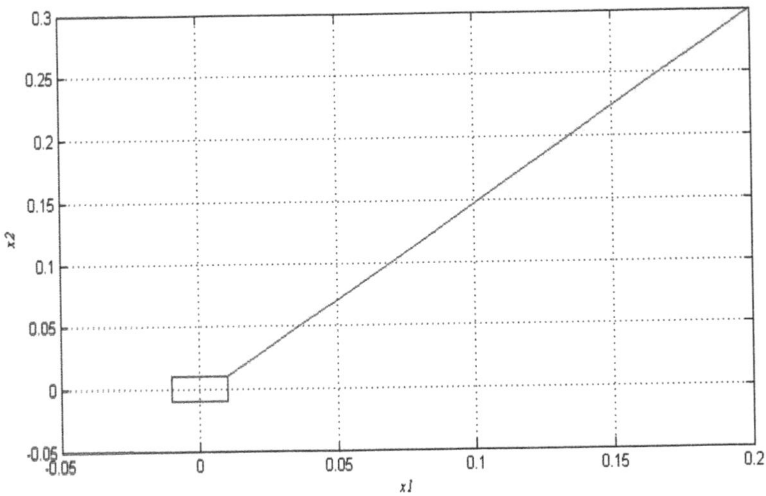

Figure C.1. *Evolution of system variables to the attractor*

Appendix D

Overvaluation of the Maximum Error in a Tracking Problem for a Lur'e Postnikov Type Process

Let us consider the third-order Lur'e Postnikov type system with the state vector x_{I1} and the system with uncertainties of state vector x_{I2}, such that

$$\begin{cases} \dot{x}_{I2} = A_I x_{I2} + B_{I2} f(\varepsilon_2) + \delta_I(.) \\ \varepsilon_2 = e - C_I^T x_{I2} \end{cases} \quad \text{[D.1]}$$

with

$$A_I = \begin{bmatrix} 2.25 & -10.75 & -1.375 \\ 3.5 & -10 & -0.75 \\ -39.5 & 72.5 & 5.25 \end{bmatrix}; \quad B_I = \begin{bmatrix} 23.25 \\ 13.5 \\ -23.5 \end{bmatrix}; \quad C_I^T = \begin{bmatrix} 1 & -2 & -0.5 \end{bmatrix}. \quad \text{[D.2]}$$

First, we must carry out a change of basis defined by the matrix P_I so that $C_I^T x_{I2}$ is the last component of the state vector x; for the vector characterizing the output, we obtain:

$$x_{I2} = P_I x_2 \quad \text{[D.3]}$$

$$C^T = C_I^T P_I = \begin{bmatrix} 0 & 0 & 1 \end{bmatrix}. \quad \text{[D.4]}$$

There are an infinite number of basis changes that give this result.

For P_I defined by

$$P_I = \begin{bmatrix} -2 & 0.5 & 0 \\ -1 & 0 & 0 \\ 0 & 1 & -2 \end{bmatrix} \quad [D.5]$$

we obtain

$$\dot{x}_2 = Ax_2 + Bf(\varepsilon_2) + \delta(.) \quad [D.6]$$

with

$$A = P_I^{-1} A_I P_I; \quad B = P_I^{-1} B_I; \quad \delta(.) = P_I^{-1} \delta_I(.) \quad [D.7]$$

and

$$A = \begin{bmatrix} -3 & -1 & -1.5 \\ 0.5 & -4.5 & -0.5 \\ -3 & 5 & 5 \end{bmatrix}; \quad B = \begin{bmatrix} -13.5 \\ -7.5 \\ 8 \end{bmatrix}; \quad C^T = \begin{bmatrix} 0 & 0 & 1 \end{bmatrix} \quad [D.8]$$

or for the process without input:

$$\dot{x}_2 = (A - Bf^* C^T) x_2 + \delta(.)$$

$$\dot{x}_2 = \psi(x_2) x_2 + \delta(.) \quad [D.9]$$

with

$$\psi(.) = A - Bf^*(.) C^T \quad [D.10]$$

by letting

$$f(\varepsilon_2) = \frac{f(\varepsilon_2)}{\varepsilon_2} \varepsilon_2 = f^* \varepsilon_2 . \quad [D.11]$$

The change of basis defined by matrix P

$$P=\begin{bmatrix}2 & 1 & 0\\ 1 & 1 & 0\\ 0 & 0 & 1\end{bmatrix} \quad \text{[D.12]}$$

leads to a characterization of the overall system through the characteristic matrix $D_{\psi}(x_2)=P^{-1}\psi(x_2)P$ in arrow form:

$$\dot{x}_{2D}=P^{-1}APx_{2D}-P^{-1}BC^{T}f^{*}Px_{2D}+P^{-1}\delta(.) \quad \text{[D.13]}$$

$$\dot{x}_{2D}=D_{\psi}(x_2)x_{2D}+\delta_D(.) \quad \text{[D.14]}$$

with

$$D_{\psi}(x_2)=\begin{bmatrix}-3.5 & 0 & -1+6f^{*}\\ 0 & -4 & 0.5+1.5f^{*}\\ -1 & 2 & 5-8f^{*}\end{bmatrix} \quad \text{[D.15]}$$

Study on stability and determining the attractor using aggregation techniques

The vector norm $p(x_{2D})=\left[\left|x_{2D_1}\right|,\left|x_{2D_2}\right|\ldots,\left|x_{2D_n}\right|\right]^{T}$ allows an overvaluation $M(f^{*}(.))$ of matrix $D_{\psi}(x_2)$.

$$\frac{dp}{dt}\left|x_{2D}\right|\leq M(f^{*})p(\left|x_{2D}\right|)+\left|\delta_D(.)\right| \quad \text{[D.16]}$$

with

$$M(f^{*})=\begin{bmatrix}-3.5 & 0 & \left|-1+6f^{*}\right|\\ 0 & -4 & \left|0.5+1.5f^{*}\right|\\ 1 & 2 & 5-8f^{*}\end{bmatrix}. \quad \text{[D.17]}$$

The stability of matrix $M(f^*)$ imposes that $f^* > 0.8968$.

We get the comparison system corresponding to the least favorable conditions:

$$\dot{z} = Mz + N \tag{D.18}$$

with

$$M = \begin{bmatrix} -3.5 & 0 & \left|-1+6f^*\right| \\ 0 & -4 & \left|0.5+1.5f^*\right| \\ 1 & 2 & \left(5-8f^*\right) \end{bmatrix}. \tag{D.19}$$

Let us consider the case where:

$$2 \le f^* \le f_M . \tag{D.20}$$

For the least favorable case corresponding to $f^* = f_m^* = 2$, we obtain

$$M = \begin{bmatrix} -3.5 & 0 & 11 \\ 0 & -4 & 3.5 \\ 1 & 2 & -11 \end{bmatrix} \tag{D.21}$$

Since $\delta_D(.) = P^{-1}P_I^{-1}\delta_I(.)$, we obtain:

$$N = \max\left|\delta_D(.)\right| = \max\left|P^{-1}P_I^{-1}\delta_I(.)\right| . \tag{D.22}$$

Let

$$\delta_{\mathrm{I m}} \prec \delta_I \prec \delta_{\mathrm{I}M} \tag{D.23}$$

with

$$\delta_{\mathrm{I m}} = \begin{bmatrix} -0.1 & -0.05 & -0.15 \end{bmatrix}^T ; \quad \delta_{\mathrm{I}M} = \begin{bmatrix} 0.05 & 0.1 & 0.1 \end{bmatrix}^T \tag{D.24}$$

thus

$$P^{-1}P_I^{-1}\delta_{\mathrm{Im}} \prec \delta_D \prec P^{-1}P_I^{-1}\delta_{\mathrm{I}M} \qquad \text{[D.25]}$$

with

$$P^{-1}P_I^{-1} = \begin{bmatrix} -2 & 3 & 0 \\ 4 & -7 & 0 \\ 1 & -2 & -0.5 \end{bmatrix} \qquad \text{[D.26]}$$

we have

$$\begin{bmatrix} 0.05 & -0.05 & 0.075 \end{bmatrix}^T \prec \delta_D \prec \begin{bmatrix} 0.2 & -0.5 & -0.2 \end{bmatrix}^T \qquad \text{[D.27]}$$

and

$$N = \max\left|\delta_D(.)\right| = \begin{bmatrix} 0.2 & 0.5 & 0.2 \end{bmatrix}^T \qquad \text{[D.28]}$$

If $M(D_\psi(x_2))$ is the opposite of an M-matrix, we get:

$$\lim_{t\to+\infty} p(x_{2D}) \le \lim_{t\to+\infty} z = -M^{-1}N \qquad \text{[D.29]}$$

or in the initial representation

$$\lim_{t\to+\infty}\left|P^{-1}P_I^{-1}x_{I2}\right| \le -M^{-1}N \qquad \text{[D.30]}$$

meaning that

$$\lim_{t\to\infty} p(P^{-1}P_I^{-1}x_{I2}) \le \begin{bmatrix} 0.3181 \\ 0.1977 \\ 0.0830 \end{bmatrix}. \qquad \text{[D.31]}$$

Figure D.1 shows the evolution of the state of the initial system in autonomous mode toward the previously calculated attractor.

Let us now consider a case where input e is non-zero

Let us consider the process of a system without uncertainties for which the initial model corresponds to state vector x_{I1}.

$$\begin{cases} \dot{x}_{I1} = Ax_{I1} + Bf(\varepsilon_1) \\ \varepsilon_1 = e - C^T x_{I1} \end{cases} \qquad \text{[D.32]}$$

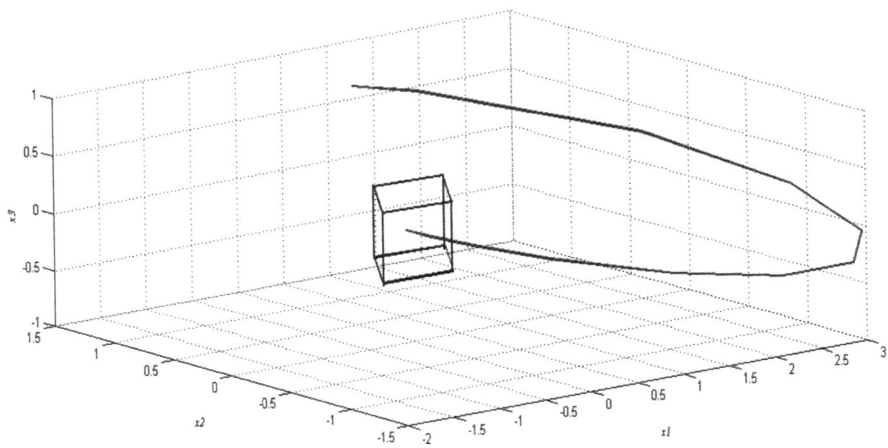

Figure D.1. *Evolution of the state of the initial system in autonomous mode toward its attractor*

The evolution of the deviation of responses from the nominal system of state vector x_{I1} and from the uncertain system subjected to the same input state vector x_{I2} is described by the following equation:

$$\begin{cases} \dot{x}_{I1} - \dot{x}_{I2} = A_I(x_{I1} - x_{I2}) + B_I(f(\varepsilon_1) - f(\varepsilon_2)) - \delta(.) \\ \varepsilon_1 - \varepsilon_2 = -C_I^T(x_{I1} - x_{I2}) \end{cases} \qquad \text{[D.33]}$$

If f(.) is differentiable, applying the theorem of finite increments gives:

$$f(\varepsilon_1) - f(\varepsilon_2) = f^*(\theta)(\varepsilon_1 - \varepsilon_2) \qquad \text{[D.34]}$$

with $\varepsilon_1 \prec \theta \prec \varepsilon_2$

we obtain

$$\dot{x}_{I1} - \dot{x}_{I2} = A_I(x_{I1} - x_{I2}) - B_I f^*(\varepsilon) C_I^T (x_{I1} - x_{I2}) - \delta(.) \qquad [D.35]$$

thus

$$x_{I1} - x_{I2} = \xi_I \ . \qquad [D.36]$$

We intend to first apply a basis change P_I such that $C_I^T \xi_I$ corresponds to the last component of the state vector ζ ; $\xi_I = P_I \xi$. In this case, we obtain:

$$P_I^{-1}\dot{\xi}_I = P_I^{-1} A_I P_I P_I^{-1} \xi_I - P_I^{-1} B_I f^*(.) C_I^T P_I P_I^{-1} \xi_I - P_I^{-1} \delta_I(.) \qquad [D.37]$$

so

$$\dot{\xi} = \left(A - Bf^*(.)C^T \right) \xi - \delta(.) \qquad [D.38]$$

of the form:

$$\dot{\xi} = \varphi(.)\xi - \delta(.) \qquad [D.39]$$

with

$$\varphi(.) = A - BC^T f^*(.) \ . \qquad [D.40]$$

We then apply the basis change P, which leads to the characteristic arrow-form matrix, we obtain:

$$P^{-1}\dot{\xi} = P^{-1} APP^{-1}\xi - P^{-1}Bf^*(\varepsilon)C^T PP^{-1}\xi - P^{-1}\delta(.) \qquad [D.41]$$

or even

$$\dot{\xi}_D = \left(A_D - B_D f^*(.) C_D^T \right) \xi_D - \delta_D(.) \qquad [D.42]$$

hence

$$\dot{\xi}_D = \begin{bmatrix} -3.5 & 0 & -1+6f^* \\ 0 & -4 & 0.5+1.5f^* \\ -1 & 2 & 5-8f^* \end{bmatrix} \xi_D - \delta_D(.) \ . \qquad [D.43]$$

Overvaluation of matrix $\left(A_D - B_D f^*(.)C_D^T\right)$ is defined as above by:

$$M\left(f^*\right) = \begin{bmatrix} -3.5 & 0 & \left|-1+6f^*\right| \\ 0 & -4 & \left|0.5+1.5f^*\right| \\ 1 & 2 & \left(5-8f^*\right) \end{bmatrix} \tag{D.44}$$

Let us consider a nonlinearity f (.) such that

$$f_m^* \leq f^* \leq f_M^* \tag{D.45}$$

If, for example, $f_m^* = 1.5$ and $f_M^* = 2$, in light of the preceding results relating to the least favorable case, we obtain:

$$M = \begin{bmatrix} -3.5 & 0 & \left|-1+6f_M^*\right| \\ 0 & -4 & \left|0.5+1.5f_M^*\right| \\ 1 & 2 & \left(5-8f_m^*\right) \end{bmatrix} \tag{D.46}$$

which is expressed digitally by

$$M = \begin{bmatrix} -3.5 & 0 & 11 \\ 0 & -4 & 3.5 \\ 1 & 2 & -7 \end{bmatrix} \tag{D.47}$$

Since the stability conditions are satisfied, matrix M is the opposite of an M-matrix, and the system is stable.

According to equation [D.22], N is defined by:

$$N = \max\left|\delta_D(.)\right| = \begin{bmatrix} 0.2 & 0.5 & 0.2 \end{bmatrix}^T \tag{D.48}$$

with M being the opposite of an M-matrix, we obtain:

$$\lim_{t\to+\infty} p(\xi_D) \leq \lim_{t\to+\infty} z = -M^{-1}N \tag{D.49}$$

or in the initial representation:

$$\lim_{t \to +\infty} \left| P^{-1} P_I^{-1} \xi_I \right| \leq -M^{-1} N \qquad \text{[D.50]}$$

and

$$\lim_{t \to \infty} p(P^{-1} P_I^{-1} \xi_I) \leq \begin{bmatrix} 0.8136 \\ 0.3356 \\ 0.2407 \end{bmatrix}. \qquad \text{[D.51]}$$

After a brief transient state that depends on the initial conditions, the trajectory error remains confined to the domain defined by equation [D.51].

Bibliography

[AND 71] ANDERSON B.D.O., MOORE J.B., *Linear Optimal Control*, Prentice Hall, Englewood Cliffs, 1971.

[AND 98] ANDERSON B.D.O, "From Youla Kucera to identification, adaptive and nonlinear systems", *Automatica*, vol. 34, pp. 1485–1506, 1998.

[AST 93] ASTROM K.J., "Matching criteria for control and identification", *Proceedings of European Control Conference*, vol. 3, pp. 248–251, 1993.

[AST 94] ASTROM K.J., "A new Smith predictor for controlling a process with an integrator and long dead-time", *IEEE Transaction on Automatic Control*, vol. 39, no. 2, pp. 334–345, 1994.

[AST 95] ASTROM K.J., WITTENMARK B., *Adaptive Control*, 2nd ed., Prentice Hall, Englewood Cliffs and Addison Wesley, Boston, 1995.

[AST 97] ASTROM K.J., WITTENMARK B., *Computer Controlled Systems, Theory and Design*, 3rd ed., Prentice Hall, Englewood Cliffs, 1997.

[AVO 00] AVOY M.T., JUONELA S.L.J., PATTON R. *et al.*, "IFAC Milestone Report on Industrial Applications", *IFAC Proceedings Volume*, 2000.

[BES 95] BESANCON VODA A., LANDAU I.D., "An iterative method for the autocalibration of digital controllers", *Proceeding of European Control Conference*, vol. 4a, pp. 2463–2469, 1995.

[BES 98] BESANCON–VODA A., "Iterative autocalibration of digital controllers – methodology and applications", *Control Engineering Practice*, vol. 6, no. 3, pp. 345–358, 1998.

[BIT 93] BITMEAD B., "Interactive control design approaches", *IFAC World Congress*, 1993.

[BOR 74] BORNE P., GENTINA J.C., "On the stability of large nonlinear systems", *Joint Automatic Control Conference*, Austin, 1974.

[BOR 87] BORNE P., "Nonlinear system stability. Vector norm approach", *Systems and Control Encyclopedia*, vol. 5, pp. 3402–3406, 1987.

[BOR 90] BORNE P., DAUPHIN-TANGUY G., RICHARD J.P. *et al.*, *Commande et Optimisation des Processus*, Technip, 1990.

[BOR 93a] BORNE P., DAUPHIN-TANGUY G., RICHARD J.P. *et al.*, *Analyse et Régulation des Processus Industriels, Tome1 régulation continue, Tome 2, Régulation Numérique*, Technip, 1993.

[BOR 93b] BORNE P., DAUPHIN-TANGUY G., RICHARD J.P. *et al.*, *Analyse et régulation des processus industriels*, Technip, 1993.

[BOR 98] BORNE P., ROZINOER J., DIEULOT J.Y. *et al.*, *Introduction à la commande floue*, Technip, 1998.

[BOR 07] BORNE P., BENREJEB M., HAGGEGE J., *Les réseaux de neurones, présentation et applications*, Technip, 2007.

[BOR 11] BORNE P., POPESCU D., BENREJEB M. *et al.*, *Automatique Avancée et Informatique Appliquée,* Editions de l'Académie, 2011.

[BOR 13a] BORNE P., POPESCU D., FILIP F. *et al.*, *Optimisation en Sciences de l'Ingénieur – Méthodes Exactes*, Hermès-Lavoisier, Paris, 2013.

[BOR 13b] BORNE P., POPESCU D., FILIP F. *et al.*, *Optimization in Engineering Sciences – Exact Methods*, ISTE, London, and John Wiley & Sons, New York, 2013.

[CAL 79] CALIN S., TERTISCO M., DUMITRACHE I. *et al.*, *Optimizari in automatizari industriale*, Editura Tehnica, Bucharest, 1979.

[CAM 04] CAMACHO E.F., BORDONS C., *Model Predictive Control*, 2nd ed., Springer, New York, 2004.

[CHA 02a] CHADLI M., MAQUIN D., RAGOT J., "Output stabilization in multiple model approach", *International Conference on Control Applications*, vol. 2, pp. 1315–1320, 2002.

[CHA 02b] CHADLI M., Stabilité et stabilisation des multimodèles, PhD Thesis, INPL-CRAN, Nancy, 2002.

[CHA 12] CHADLI M., BORNE P., *Multimodèles en automatique, outils avancés d'analyse et de synthèse*, Hermès-Lavoisier, Paris, 2012.

[COR 13] CORNOIU M., BARA C., POPESCU D., "Metaheuristic approcah in nonlinear systems", *UPB Scientific Bulletin, Series A*, Bucharest, 2013.

[DAH 13] DAHMANI H., PAGÈS O., EL HAJJAJI A. *et al.*, *Observer-Based Multi-Actuator Vehicle Chassis Control in Critical Situations*, IEEE, Villeneuve d'Ascq, pp. 156–161, 2013.

[DAU 04] DAUPHIN-TANGUY G., FOULOY L., POPESCU D. (eds), *Modelisation, identification et commande des systèmes*, Éditions de l'Académie roumaine, Bucharest, 2004.

[DEL 96] DELMOTE.F., DUBOIS.L., BORNE P., "A general scheme for multi-model controller using trust", *Mathematics and Computers in Simulation*, vol. 41, pp. 173–186, 1996.

[DIO 97] DION J.M., POPESCU D., *Optimisation des Systemes-Commande Optimale,* Diderot, Paris, 1997.

[DOY 92] DOYLE J.C., FRANCIS B.A., TANENBAUM A.R., *Feedback Control Theory*, MacMillan, New York, 1992.

[DUG 91] DUGARD L., LANDAU I.D., *Commande Adaptative-Metodologie et Applications*, Hermes, Paris, 1991.

[ELF 08] EL FELLY N., DIEULOT J.Y., BORNE P., "A neural approach of multiple model representation of complex processes", *International Journal of Computers, Communications & Control*, vol. 3, no. 2, pp. 39–50, 2008.

[FIL 02] FILIPESCU A., STAMATESCU S., *Systems Theory*, Matrix Rom, Bucharest 2002.

[FIL 09] FILIP F., POPESCU D., MATEESCU M., "Optimal decisions for complex systems-software packages", *Mathematics and Computers in Simulation*, vol. 76, pp. 422–429, 2009.

[FLA 94] FLAUS J.M., *La régulation industrielle*, Hermès, Paris, 1994.

[FOU 87] FOULARD C., GENTIL S., SANDRAZ J.P., *Commande et Régulation par Calculateur Numérique,* Eyrolles, Paris, 1987.

[FRA 72] FRANKS R.G.E., *Modeling and Simulation in Chemical Engineering,* John Wiley & Sons, New York, 1972.

[FRA 86] FRANKLIN G., POWELL J.D., *Feedback Control of Dynamic Systems*, Addison Wesley, Reading, 1986.

[FRA 90] FRANKLIN G.F., POWELL J.D., WORKMAN M.L., *Digital Control of Dynamic Systems*, Addison Wesley, Reading, London, 1990.

[GEN 97] GENTIL S., POPESCU D., *Commande Numérique et Intelligence Artificielle en Automatique*, Editura Tehnica, Bucarest, 1997.

[GEV 95] GEVERS M., "Identification for control", *Proceedings of IFAC Conference ACASP'* 95, Budapest, pp. 1–12, 1995.

[GHA 13] GHARBI A., PETRESCU C., BENREJEB M. *et al.*, "New approach for the determination of attractors for nonlinear systems", *Proceedings of 2nd International Conference on Systems and Computer Science (ICSCS 2013)*, Villeneuve d'Ascq, France, pp. 216–221, 2013.

[GHA 14] GHARBI A., BENREJEB M., BORNE P., "Tracking error estimation of uncertain Lur'e Postnikov Systems", *IEEE International Conference on Control, Decision and Information Technologies (CODIT 2014)*, Metz, France, 2014.

[GHA 15a] GHARBI A., BENREJEB M., BORNE P., "Determination of nested attractors for uncertain nonlinear systems", *IEEE International Conference CESA 2015*, Marrakech, 2015.

[GHA 15b] GHARBI A., BENREJEB M., BORNE P., "Error estimation in the decoupling of ill-defined and/or perturbed nonlinear processes", *Proceedings of 19th International Conference on Circuits, Systems, Communications and Computers (CSCC 2015)*, pp. 203–208, 2015.

[GHA 16a] GHARBI A., BENREJEB M., BORNE P., "A taboo search optimization of the control low of the bounded uncertainties nonlinear systems", *International Journal of Computers Communication and Control*, vol. 2. 2016.

[GHA 16b] GHARBI A., BENREJEB M., "Study of the stabilization of uncertain nonlinear systems controlled by state feedback", *Acta Polytechnica Hungaria*, vol. 13, no. 4, pp. 21–38, 2016.

[GHE 07] GHEDIRA K., *Optimization combinatoire par métaheuristiques*, Technip, Paris, 2007.

[GOO 84] GOODWIN G.C., SIN K.S., *Adaptive Filtering Prediction and Control*, Prentice Hall, Englewood Cliffs, 1984.

[GRU 04] GRUYITCH LJ.T., RICHARD J.P., BORNE P. *et al.*, *Stability Domains*, Chapman and Hall, 2004.

[HAB 99] HABER R., Keviczky L., *Nonlinear System Identification – Input-Output Modeling Approach*, Kluwer Academic Publishers, Dordrecht, 1999.

[HED 05] HEDRICK J.K., GIRARD A., "Feedback linearization", Control of Nonlinear Dynamic Systems: Theory and Applications, Printed Course, Universita Degli Studi Di Padova, 2005.

[HUA 04] HUANG J., *Nonlinear Output Regulation – Theory and Applications*, SIAM Advances in Design and Control Series, Philadelphia, 2004.

[KAI 80] KAILATH T., *Linear Systems*, Prentice Hall, Englewood Cliffs, 1980.

[KAR 00] KARNOPP D., MARGOLIS D., ROSENBERG R., *System Dynamics: Modelling and Simulation of Mechatronic Systems*, John Wiley & Sons, London, 2000.

[KAR 91] KARNOPP D., ROSENBERG R., *Systems Dynamics: A Unified Approach*, 2nd ed., John Wiley & Sons, London, 1991.

[KUO 91] KUO B.C., *Automatic Control Systems*, Prentice Hall, 1991.

[LAN 86] LANDAU I.D., DUGARD L., *Commande Adaptative Aspects Pratiques et Theoriques*, Masson, Paris, 1986.

[LAN 89] LANDAU I.D., "Adaptive control, robustness and performance enhancement", *IFAC International Symposium*, Glasgow, 1989.

[LAN 93] LANDAU I.D., *Identification et Commande des Systèmes*, Hermes, Paris, 1993.

[LAN 95] LANDAU I.D., *Identification et Commande des Systèmes*, Hermes, Paris, 1995.

[LAN 97] LANDAU I.D., LOZANO R., M'SAAD M., *Adaptive Control*, Springer-Verlag, London, 1997.

[LAN 01] LANDAU I.D., KARIMI A., "Identification des modèles de procédé en boucle fermée", in LANDAU I.D., BESANÇON-VODA A. (eds), *Identification de systèmes*, Hermès, Paris, 2001.

[LAN 05a] LANDAU I.D., ZITO G., *Digital Control Systems – Design, Identification and Implementations*, Springer-Verlag, London, 2005.

[LAN 05b] LANDAU I.D., CONSTANTINESCU A., REY D., "Adaptive narrow band disturbance rejection applied to an active suspension – an internal model principle approach", *Automatica*, vol. 41, pp. 563–574, 2005.

[LEV 04] LEVINE J., Analyse et Commande des Systèmes Non linéaires, Centre Automatique et Systèmes, Ecole de Mines de Paris, Paris, 2004.

[LJU 99] LJUNG L., *System Identification – Theory for the User*, 2nd ed., Prentice Hall, Upper Saddle River, 1999.

[LUP 13] LUPU C., BANDICI L., MOLNAR C. *et al.*, "Reduced simulation strategy-multiple model applicability", *Journal of Electrical and Electronics Engineering*, vol. 6, no. 1, pp. 63–68, 2013.

[M'SA 94] M'SAAD M., HEJDA I., "Adaptive control of a flexible transmission system", *Control Engineering Practice*, vol. 4, pp. 629–639, 1994.

[MAC 01] MACEJOWSKI J.M., *Predictive Control with Constraints*, Prentice Hall, 2001.

[MAM 75] MAMDANI E.H., ASSILIAN S., "An experiment in linguistic synthesis with fuzzy logic controller", *International Journal of Man-Machine Studies*, vol. 7, no. 1, pp. 1–3, 1975.

[MED 13] MEDIANU S.O., LUPU C., POPESCU D. *et al.*, *Optimal Closed-Loop Identification Solution Using the Direct and Indirect Methods for a Flexible Transmission System*, IEEE, Villeneuve d'Ascq, France, pp. 74–79, 2013.

[MED 14] MEDIANU S., LEFEVRE L., POPESCU D. *et al.*, "Simulator for open loop identification of dynamic systems", *UPB Scientific Bulletin, Series C*, Bucharest, 2014.

[MED 16] MEDIANU S., LUPU C., LEFEVRE L. *et al.*, "Solution for model adequacy verification in indirect adaptive control", *Control Engineering and Applied Informatics Journal*, vol. 18, no. 1, Bucharest, 2016.

[MIR 15] MIRON C., POPESCU D., CHRISTOV N. *et al.*, "Control based observer for PV system by fuzzy Takagi Sugeno model", *ICSTCC 18*, Sinaia, 2015.

[MIR 16] MIRON C., OLTEANU S., CHRISTOV N., "Energy management for PV systems using fuel cells", *ICSTCC 19*, Sinaia, 2016.

[MOR 00] MORERE. Y., Mise en œuvre de lois de commande pour les modèles flous de type Takagi-Sugeno", PdD Thesis, University of Valenciennes and Harnaut Cambiésis, 2000.

[OGA 91] OGATA K., *Modern Control Engineering*, Prentice Hall, 1990.

[OPP 85a] OPPENHEIM A.V., SCHAFER R., *Digital Signal Processing*, Prentice Hall, Englewood Cliffs, New York, 1985.

[OPP 85b] OPPENHEIM A.V., WILLSKY A.S., *Signals and Systems*, Prentice Hall, Englewood Cliffs, 1985.

[POP 98] POPESCU D., STEFANOIU D., "System identification and command", Editura Tehnica, Bucharest, February 1998.

[POP 00] POPESCU D., TAMBOZI G., PETRESCU C., "Some considerations about the rafinament and accuracy of PID controllers", *Proceedings of IFAC Workshop on Digital Control*, Terrasa, 2000.

[POP 01a] POPESCU D., SERBANESCU M., "Software package for optimal decisions large scale systems", *IFAC Symposium on Large Scale Systems*, LSS Bucharest, pp. 225–230, 2001.

[POP 01b] POPESCU D., FANEA A., "Refinement and accuracy of PID algorithms", *IFAC Symposium on System Structure and Control*, SSC 2001, Prague, 2001.

[POP 11a] POPESCU D., IONESCU F., MIRCIOIU M. *et al.*, *Robust Control and Optimization for an Ethylene Reactor*, 18th Mediterranean Conference on Control and Automation (MED), pp. 1097–1102, Marrakech, 2011.

[POP 11b] POPESCU D., IONESCU F., DOBRESCU R. *et al.*, *Modeling in Engineering Industrial Processes*, Ed. AGIR Press, Bucharest, 2011.

[POP 13] POPESCU D., PETRESCU C., DIMON C. *et al.*, *Control and Optimization for a Petrochemical Reactor*, IEEE, Villeneuve d'Ascq, pp. 14–19, 2013.

[ROB 01] ROBERTS P.D., "Optimal control using integrated system optimization and parameter estimation", *Proceedings of IFAC Symposium on Large Scale Systems*, LSS 2001, Bucharest, pp. 1–13, 2001.

[ROS 83] ROSENBERG R., KARNOPP D., *Introduction to Physical System Dynamics*, McGraw-Hill, New York, 1983.

[SHE 50] SHERMAN J., MORRISON W.J., "Adjustment of an inverse matrix corresponding to a change in one element of a given matrix", *Annals of Mathematical Statistics*, vol. 21, no. 1, pp. 124–127, 1950.

[SÖD 89] SÖDERSTRÖM T., STOICA P., *System Identification*, Prentice Hall, London, 1989.

[STE 05] STEFANOIU D., CULITA J., STOICA P., *A Foundation to System Modeling and Identification*, Printech Press, Bucharest, 2005.

[STE 10] STEFANOIU D., LANDAU I.D., CORNOIU M., "Identification of an active suspension in the presence of narrow band disturbances", *IFAC Conference on Management and Control of Production Logistics – MCPL-2010*, Coimbra, paper 63, 2010.

[STE 14] STEFANOIU D., BORNE P., POPESCU D. *et al.*, "Metaheuristics – local methods", in *Optimisation in Engineering Sciences. Metaheuristics, Stochastic Methods and Decision Support*, ISTE, London and John Wiley & Sons, New York, 2014.

[STE 16] STEFANOIU D., BARRIO A.M., STOICA A.M. *et al.*, "*A RST* design approach for the launchers flight control system", *The 6-th International Conference on Astrodynamics Tools and Techniques – ICATT-2016*, Darmstadt, pp. 32–37, 2016.

[TAK 85] TAKAGI M., SUGENO M., "Fuzzy identification of systems and its application to modelling and control", *IEEE Transactions on Systems, Man and Cybernetics*, vol. 15, no. 1, pp. 116–132, 1985.

[TAL 93] TALOI D., *Optimization of Metallurgical Processes*, Ed.Tehnica, Bucharest, 1993.

[TAN 00] TANGUY G.D., *Les bond graphs*, Hermès, Paris, 2000.

[TAN 13] TANASA V., CHIRIAC A., POPESCU D. *et al.*, *Errors Estimation for the Numerical Control Computations in the Industrial Applications*, ICSTCC, Sinaia, 2013.

[TAN 99] TANGUY G.D., *Les bond graphs et leur application en mécatronique*, Techniques de l'Ingénieur, Paris, 1999.

[TER 98] TERTISCO M., POPESCU D., JORA B. *et al.*, *Automatizari industriale continue*, Editura Didactica si Pedagogica, Bucharest, 1998.

[THO 75] THOMA J., *Introduction to Bond Graphs and Their Applications*, Pergamon Press, Oxford, 1975.

[TOR 15] TOROUS C., POPESCU D., BALAN D. *et al.*, *Search Algorithms for the Maximum Power Point on Photovoltaic Panels*, ICSTCC 18, Sinaia, 2015.

[TOR 16] TOROUS C., OLTEANU S., POPESCU D., *Extremal Control for Photovoltaic Panels,* ICSTCC 19, Sinaia, 2016

[TUD 13] TUDOR F., POPESCU D., BORNE P., *Advanced Control for Thermo-Energetic Installations*, ICSTCC, Sinaia, 2013.

[VAN 95] VAN DER HOP P., SCHRAMA R., "Identification and control", *Automatica*, vol. 31, no. 12, pp. 1751–1770, 1995.

[WAN 09] WANG L., *Model Predictive Control, System Design and Implementation Using Matlab*, Springer-Verlag, London, 2009.

[WEI 62] WEIGHT W.B., SCRIMGEOUR J.H.C., *Modeling Steel Mills on a Digital Computer*, Iron and Steel Engineering, 1962.

[WIR 91] WIRK G.S., *Digital Computer Systems*, MacMillan, London, 1991.

[WIS 72] WISMER D.A. *Optimization Methods for Large Seale Systems with Applications*, McGrow-Hill Inc., New York, 1972.

[XU 93] XU L., KROYZAK A., ORIA E., "Rival penalized competitive learning for cluster analysis RBF and curve detection", *IEEE Transactions on Neural Network*, vol. 4, pp. 636–646, 1993.

[ZAD 65] ZADEH L., "Fuzzy sets", *Information Control*, vol. 8, pp. 338–353, 1965.

Index

P

R, S, T, U

Other titles from

iSTE

in

Systems and Industrial Engineering – Robotics

2017

FEYEL Philippe
Robust Control, Optimization with Metaheuristics

MARÉ Jean-Charles
Aerospace Actuators 2: Signal-by-Wire and Power-by-Wire

RÉVEILLAC Jean-Michel
Modeling and Simulation of Logistics Flows 1: Theory and Fundamentals
Modeling and Simulation of Logistics Flows 2: Dashboards, Traffic Planning and Management
Modeling and Simulation of Logistics Flows 3: Discrete and Continuous Flows in 2D/3D

2016

ANDRÉ Michel, SAMARAS Zissis
Energy and Environment
(Research for Innovative Transports Set - Volume 1)

AUBRY Jean-François, BRINZEI Nicolae, MAZOUNI Mohammed-Habib
Systems Dependability Assessment: Benefits of Petri Net Models (Systems Dependability Assessment Set - Volume 1)

BLANQUART Corinne, CLAUSEN Uwe, JACOB Bernard
Towards Innovative Freight and Logistics (Research for Innovative Transports Set - Volume 2)

COHEN Simon, YANNIS George
Traffic Management (Research for Innovative Transports Set - Volume 3)

MARÉ Jean-Charles
Aerospace Actuators 1: Needs, Reliability and Hydraulic Power Solutions

REZG Nidhal, HAJEJ Zied, BOSCHIAN-CAMPANER Valerio
Production and Maintenance Optimization Problems: Logistic Constraints and Leasing Warranty Services

TORRENTI Jean-Michel, LA TORRE Francesca
Materials and Infrastructures 1 (Research for Innovative Transports Set - Volume 5A)
Materials and Infrastructures 2 (Research for Innovative Transports Set - Volume 5B)

WEBER Philippe, SIMON Christophe
Benefits of Bayesian Network Models
(Systems Dependability Assessment Set – Volume 2)

YANNIS George, COHEN Simon
Traffic Safety (Research for Innovative Transports Set - Volume 4)

2015

AUBRY Jean-François, BRINZEI Nicolae
Systems Dependability Assessment: Modeling with Graphs and Finite State Automata

BOULANGER Jean-Louis
CENELEC 50128 and IEC 62279 Standards

BRIFFAUT Jean-Pierre
E-Enabled Operations Management

MISSIKOFF Michele, CANDUCCI Massimo, MAIDEN Neil
Enterprise Innovation

2014

CHETTO Maryline
Real-time Systems Scheduling
Volume 1 – Fundamentals
Volume 2 – Focuses

DAVIM J. Paulo
Machinability of Advanced Materials

ESTAMPE Dominique
Supply Chain Performance and Evaluation Models

FAVRE Bernard
Introduction to Sustainable Transports

GAUTHIER Michaël, ANDREFF Nicolas, DOMBRE Etienne
Intracorporeal Robotics: From Milliscale to Nanoscale

MICOUIN Patrice
Model Based Systems Engineering: Fundamentals and Methods

MILLOT Patrick
Designing Human–Machine Cooperation Systems

NI Zhenjiang, PACORET Céline, BENOSMAN Ryad, REGNIER Stéphane
Haptic Feedback Teleoperation of Optical Tweezers

OUSTALOUP Alain
Diversity and Non-integer Differentiation for System Dynamics

REZG Nidhal, DELLAGI Sofien, KHATAD Abdelhakim
Joint Optimization of Maintenance and Production Policies

STEFANOIU Dan, BORNE Pierre, POPESCU Dumitru, FILIP Florin Gh., EL KAMEL Abdelkader
Optimization in Engineering Sciences: Metaheuristics, Stochastic Methods and Decision Support

2013

ALAZARD Daniel
Reverse Engineering in Control Design

ARIOUI Hichem, NEHAOUA Lamri
Driving Simulation

CHADLI Mohammed, COPPIER Hervé
Command-control for Real-time Systems

DAAFOUZ Jamal, TARBOURIECH Sophie, SIGALOTTI Mario
Hybrid Systems with Constraints

FEYEL Philippe
Loop-shaping Robust Control

FLAUS Jean-Marie
Risk Analysis: Socio-technical and Industrial Systems

FRIBOURG Laurent, SOULAT Romain
Control of Switching Systems by Invariance Analysis: Application to Power Electronics

GROSSARD Mathieu, REGNIER Stéphane, CHAILLET Nicolas
Flexible Robotics: Applications to Multiscale Manipulations

GRUNN Emmanuel, PHAM Anh Tuan
Modeling of Complex Systems: Application to Aeronautical Dynamics

HABIB Maki K., DAVIM J. Paulo
Interdisciplinary Mechatronics: Engineering Science and Research Development

HAMMADI Slim, KSOURI Mekki
Multimodal Transport Systems

JARBOUI Bassem, SIARRY Patrick, TEGHEM Jacques
Metaheuristics for Production Scheduling

KIRILLOV Oleg N., PELINOVSKY Dmitry E.
Nonlinear Physical Systems

LE Vu Tuan Hieu, STOICA Cristina, ALAMO Teodoro, CAMACHO Eduardo F., DUMUR Didier
Zonotopes: From Guaranteed State-estimation to Control

MACHADO Carolina, DAVIM J. Paulo
Management and Engineering Innovation

MORANA Joëlle
Sustainable Supply Chain Management

SANDOU Guillaume
Metaheuristic Optimization for the Design of Automatic Control Laws

STOICAN Florin, OLARU Sorin
Set-theoretic Fault Detection in Multisensor Systems

2012

AÏT-KADI Daoud, CHOUINARD Marc, MARCOTTE Suzanne, RIOPEL Diane
Sustainable Reverse Logistics Network: Engineering and Management

BORNE Pierre, POPESCU Dumitru, FILIP Florin G., STEFANOIU Dan
Optimization in Engineering Sciences: Exact Methods

CHADLI Mohammed, BORNE Pierre
Multiple Models Approach in Automation: Takagi-Sugeno Fuzzy Systems

DAVIM J.Paulo
Lasers in Manufacturing

DECLERCK Philippe
Discrete Event Systems in Dioid Algebra and Conventional Algebra

DOUMIATI Moustapha, CHARARA Ali, VICTORINO Alessandro, LECHNER Daniel
Vehicle Dynamics Estimation using Kalman Filtering: Experimental Validation

GUERRERO José A, LOZANO Rogelio
Flight Formation Control

HAMMADI Slim, KSOURI Mekki
Advanced Mobility and Transport Engineering

MAILLARD Pierre
Competitive Quality Strategies

MATTA Nada, VANDENBOOMGAERDE Yves, ARLAT Jean
Supervision and Safety of Complex Systems

POLER Raul *et al.*
Intelligent Non-hierarchical Manufacturing Networks

TROCCAZ Jocelyne
Medical Robotics

YALAOUI Alice, CHEHADE Hicham, YALAOUI Farouk, AMODEO Lionel
Optimization of Logistics

ZELM Martin *et al.*
Enterprise Interoperability –I-EASA12 Proceedings

2011

CANTOT Pascal, LUZEAUX Dominique
Simulation and Modeling of Systems of Systems

DAVIM J. Paulo
Mechatronics

DAVIM J. Paulo
Wood Machining

GROUS Ammar
Applied Metrology for Manufacturing Engineering

KOLSKI Christophe
Human–Computer Interactions in Transport

LUZEAUX Dominique, RUAULT Jean-René, WIPPLER Jean-Luc
Complex Systems and Systems of Systems Engineering

ZELM Martin, *et al.*
Enterprise Interoperability: IWEI2011 Proceedings

2010

BOTTA-GENOULAZ Valérie, CAMPAGNE Jean-Pierre, LLERENA Daniel, PELLEGRIN Claude
Supply Chain Performance / Collaboration, Alignement and Coordination

BOURLÈS Henri, GODFREY K.C. Kwan
Linear Systems

BOURRIERES Jean-Paul
Proceedings of CEISIE'09

CHAILLET Nicolas, REGNIER Stéphane
Microrobotics for Micromanipulation

DAVIM J. Paulo
Sustainable Manufacturing

GIORDANO Max, MATHIEU Luc, VILLENEUVE François
Product Life-Cycle Management / Geometric Variations

LOZANO Rogelio
Unmanned Aerial Vehicles / Embedded Control

LUZEAUX Dominique, RUAULT Jean-René
Systems of Systems

VILLENEUVE François, MATHIEU Luc
Geometric Tolerancing of Products

2009

DIAZ Michel
Petri Nets / Fundamental Models, Verification and Applications

OZEL Tugrul, DAVIM J. Paulo
Intelligent Machining

PITRAT Jacques
Artificial Beings

2008

ARTIGUES Christian, DEMASSEY Sophie, NERON Emmanuel
Resources–Constrained Project Scheduling

BILLAUT Jean-Charles, MOUKRIM Aziz, SANLAVILLE Eric
Flexibility and Robustness in Scheduling

DOCHAIN Denis
Bioprocess Control

LOPEZ Pierre, ROUBELLAT François
Production Scheduling

THIERRY Caroline, THOMAS André, BEL Gérard
Supply Chain Simulation and Management

2007

DE LARMINAT Philippe
Analysis and Control of Linear Systems

DOMBRE Etienne, KHALIL Wisama
Robot Manipulators

LAMNABHI Françoise *et al.*
Taming Heterogeneity and Complexity of Embedded Control

LIMNIOS Nikolaos
Fault Trees

2006

FRENCH COLLEGE OF METROLOGY
Metrology in Industry

NAJIM Kaddour
Control of Continuous Linear Systems

Printed and bound by CPI Group (UK) Ltd, Croydon, CR0 4YY
16/09/2026
14963078-0005